AF564766

BIOTECHNOLOGY OF ANIMAL CULTURE

BIOTECHNOLOGY OF ANIMAL CULTURE

By

Dr. P.R. Yadav

Lecturer

Department of Zoology

D.A.V. College

Muzaffarnagar (U.P.)

&

Dr. Rajiv Tyagi

Department of Zoology

M.M. College

Modi Nagar (U.P.)

DISCOVERY PUBLISHING HOUSE

NEW DELHI-110002

First Published-2006
Reprint: 2011
ISBN 81-8356-101-2

Published by

DISCOVERY PUBLISHING HOUSE
4831/24, Ansari Road, Prahlad Street,
Darya Ganj, New Delhi-110002 (India)
Phone: 23279245 • Fax: 91-11-23253475
E-mail:dphtemp@indiatimes.com

Printed at:
Dynamic Printers, Delhi

Preface

The present title "Biotechnology of Animal Culture" has been designed for undergraduate and post-graduate students of Indian Universities. The book is intended to meet the needs of the shorter, less encyclopedic course in introductory Biotechnology of Animal Culture. The book concentrates on the basic techniques easily applicable in the labs, omitting the encumbering details of many advanced and specialized methods. The focus is on the basic principles of this branch of Biotechnology and on explaining these fundamentals to students in as clear and concise a way as possible.

This branch of science in its diverse facets is presently advancing in meteroic speed. By the time this volume could be completed, Biotechnology may have advanced further in exciting direction.

To make the work more comprehensive and informative, the author has consulted many authoritative books, research journals, abstracts, monographs etc. He is grateful to all those great scholars whose work are cited or substantially reproduced.

There can be no claim to originality except in the manner of treatment and much of the information has been obtained from the books and scientific journals available in the different libraries.

The author expresses his thanks to his friends and colleagues whose continue inspirations have initiated him to bring out this book.

The author is painfully aware of the shortcomings, errors and misprints that have crept in, and shall be greateful to receive suggestion for improvement of the next edition from all the readers.

The author expresses his gratitude to Mr. Wasan and staff of M/s Discovery Publishing House for their whole hearted co-operation in the publication of this book.

Author

CONTENTS

1

Protozoan Culture

NATURAL POND CULTURES

These cultures should be made by the methods of Hyman, Jennings, and others. The plant material collected should not be restricted to Ceratophyllum and Elodea, but should include any vegetable matter, *e.g.*, old lily pads and stems, cat-tails (especially if decay has commenced), decaying leaves of trees, grass and sedges, *etc.*, from pools, ponds, marshes, bogs, ditches, and rivers. Chara and clean Spirogyra and the brown mat of algae from the surface of a dam or of stones are of value, if not too much is taken and if to this is added hay or other materials to furnish food for bacteria. Decaying Sphagnum from sphagnum bogs is valuable. Mud or ooze from the bottom of ponds or pools which get the drainage from pastures or barnyards is also good. In collecting this vegetable material always bring some of the water from the same situation. Most Protozoa are sensitive to changes in water. Treated water is particularly bad for some species and should never be used until it has stood for some days in an open vessel or tank.

Place natural pond cultures in shallow glass or earthenware dishes and keep covered to prevent evaporation. If pond scums and Euglena are wanted or if these are needed for the food of any desired protozoan, set the cultures in diffuse light, never in direct sunlight. Label each culture use one pipette for each culture. This precaution is of great importance in examining the subcultures.

Examination of cultures

At intervals each culture should be examined carefully. In making the examination, take samples of scum, of clear liquid, scrapings from sides of vessel, from the light side and from the dark side, from the vegetation, and from the bottom. These situations may furnish different forms because some forms swim freely while others creep. Some require much oxygen while others can exist on less.

Subcultures

If at any time any culture is yielding a large number of desirable species such as Paramecium, Euglena, Amoeba, *etc.*, subcultures should be made as follows: Clean thoroughly short tender dishes or finger bowls. Place in these filtered cistern or distilled water. Do not use raw tap water. If neither cistern nor distilled water is available filter some water which has stood for a long time in an aquarium or tank. Such water seems to have lost many of its noxious properties. Place a few straws of clean hay in the water and allow it to stand 36 to 48 hours before the desired Protozoa are placed in it. In place of dry hay, they may be boiled in distilled or cistern water and the hay, not the hay water, added to the water in the dish. Boiling the hay will sterilize it so that sterilization of the hay in an autoclave is probably unnecessary. Allow to stand before inoculating. In such cultures bacteria will develop and on these Paramecium and certain other forms feed.

Pure Cultures

Isolation of particular species is not an entirely simple matter. However, individual Protozoa may frequently be secured by using a mouth pipette. This is a glass tube with a finely drawn tip in one end of a rubber tube 12 to 15 inches long with a short piece of glass tubing to place in the mouth in the other end. Place a small dish of the culture containing the organism sought on the stage of the binocular microscope and put the glass tube in the mouth. When a desired individual is found bring the point of the pipette near it and suck on the tube. With some practice the protozoan may be caught and may be put on a slide in a small drop of water and examined. In this way individual Protozoa may be isolated for the inoculation of cultures and by washing them in one or several changes of water the worker can be reasonably sure that other species are not present.

AMOEBA

The fact is not generally appreciated that Amoebae are found in clean water, not foul. Two forms of Amoebae (*A. proteus* type) are used in our department. One of these, an unusually large and active form, possessing many pseudopodia, is found in association with catfish and newts (*Ameiurus nebulosus* and *Triturus viridescens*). Another with fewer pseudopodia and more regular in outline is taken with the Brook Stickleback (*Eucalia inconstans*). The latter form we find the more satisfactory for class work.

The Brook Sticklebacks are collected in the backwash pools of rivers or streams. Two or three of these small fish are placed in a bowl containing 2000 cc. of pond water and fed upon Daphnia or enchytraeid worms. Gradually ooze forms on the bottom of the bowl and in about two weeks this ooze will be found to contain Amoebae.

A culture of Amoebae is then set up in bowls containing 600 cc. of pond water. The bottom of each bowl is sprinkled with exceedingly fine sand which has been carefully washed and sifted through bolting cloth. The individual grains of sand should be scarcely larger, and for the most part smaller, than a single Amoeba. To each bowl is added 6 grains of boiled wheat. (The development of the culture may be hastened by using boiled brown rice in place of the wheat; although the latter culture does not last so long). Some ooze from the fish bowls is now introduced and the culture kept at about 73°F.

In order that the Amoebae may thrive, fungus must grow on the grains of wheat. If this fails to appear, it may be obtained by adding a bit of dead worm or dead fly from some other culture.

Within twelve days there should be an abundance of the Amoebae. Large numbers may be present as early as six or seven days, and usually so in eight days. A culture lasts about a month, but new ones should be started while the Amoebae are still active and healthy. Old cultures should be made over completely.

The fine sand in the culture is most helpful in picking the Amoebae out of the bowl for microscopic study, since a little sand taken into a pipette will have the organisms attached to it. The sand also forms an excellent support for the coverslip, allowing the Amoebae to move about freely. As part of the habitat of the culture the grains of sand afford a cover under which the organisms can hide. In an undisturbed culture the Amoebae will be found contracted and resting either under the grains or close beside them.

If disturbed by jarring or by a beam of light the Amoebae glide or "walk" away from this protecting cover .

Stock Cultures of *Amoeba proteus*

In the course of experiments it has been necessary to maintain cultures of *Amoeba proteus* in stock. The writer endeavored to find a medium that made requisite a minimum amount of attention. The effort in this direction met with considerable success. In view of the wide use of Amoeba of the proteus type in biological research and elementary instruction in biology, a culture medium that is simple, reproducible, and extremely reliable will be of general interest. The medium used in as follows:

NaCl	0.1 gr.
KCl	0.004 gr.
$CaCl_2$	0.006 gr.
H_2O	1000 cc. (glass distilled)

Two hundred to 250 cc. of this solution is put into a finger bowl or glass crystallizing dish 8 or 10 cm. in diameter and to each of such dishes is added 4 or 5 grains to polished rice (any brand carried at the corner grocery is suitable). The cultures thus prepared are immediately seeded with 50 to 100 Amoebae, covered with glass plates to prevent evaporation and entry of dust, and then left, preferably in a dark cool place, to develop. Such cultures will produce a fine crop in from 2 to 4 weeks and so far in some 30 to 40 cultures the writer has had only one or two failures. Out of five cultures that were set up as a test, three cultures one year old had ample numbers of Amoebae; the other two died out in eleven months.

These five cultures during their existence were deliberately neglected. No detritus was removed. Rice was added only when it was discovered that none was apparent in the culture. Water too was added to compensate for evaporation with no attempt at regularity, say, on the average of once a month. The temperature variation was from 19° to 28° C.

In other words, the cultures were subjected to as careless handling as if in the hands of a somewhat below par student assistant, but they survived.

The Culture of *Amoeba proteus*

Collection and culture. Amoeba proteus is not found in nature as frequently as are smaller Amoebae. When found it will usually be in ponds or pools where there is neither too great nor too little organic material, where there are no excessively swift currents, and in water that is not too alkaline. If, with a spoon or dipper, surface water containing a considerable amount of organic material is collected, taken to the laboratory, poured into tall slender jars and allowed to settle it will be found that the heavy organic material will settle in a thick layer on the bottom, and that the Amoebae will settle on top of this layer. They then may be removed with a pipette to a shallow dish and examined.

Amoeba proteus collected in this way may be cultured simply by placing a suitable spring or pond water in finger bowls or other shallow dishes to depth of about 2 cm., adding 3 to 4 grains of wheat, or 5 to 6 grains of polished rice, or 5 to 6 one-inch stems of timothy hay, and then inoculating with Amoebae from the collection. The Amoebae in these cultures will become very abundant in from 3 to 4 weeks. It should be noted here that Dawson gives an elaborate method for obtaining cultures from freshly collected material consisting of a gradual dilution of the collected water with distilled water. This process of dilution in our experience has not been necessary. Cultures obtained from freshly collected material are likely to contain in addition to Amoebae various organisms such as small Crustacea, rotifers, and various Protozoa. Since none of these organisms, except the cryptomonad, Chilomonas, are necessary as food, and since others such as the Crustacea probably feed on the Amoebae; it is well to take steps to eliminate these unnecessary organisms. *A. proteus* feeds on a variety of organisms, but Chilomonas alone is entirely adequate.

Freeing Amoeba cultures of contaminating organisms and establishing clone cultures. Sterilize some spring water by autoclaving for 15 minutes at 15 pounds' pressure, or merely by bringing to a boil. Allow to cool. Place sterile spring water in three or four sterile, chemically clean Syracuse watch glasses, and a finger bowl. The water in the finger bowl should have a depth of about 2 cm. Add to the finger bowl, 3 to 4 grains of wheat, 5 to 6 grains of rice, or 5 to 6 pieces of timothy hay stems about 1 inch long. The wheat, rice, or hay should be autoclaved dry 15 minutes at 15 pounds' pressure, or placed dry in a test tube, then the test tube placed in a beaker of boiling water for about 15 minutes. Now with a sterile capillary pipette and under a binocular microscope, select active Amoebae, pass them one at a time through the dishes of sterile spring water.

If it is desired to obtain a clone culture (culture containing Amoebae all of which have descended from a single parent) put only one washed Amoeba into the finger bowl culture. To obtain merely a clean culture it is advisable to put into the finger bowl culture 25 or more Amoebae. Now add to the culture in the finger bowl containing washed Amoebae a drop of culture fluid containing only Chilomonas and bacteria. The best way to do this is to take small drops of the old culture fluid, examine them carefully under high power selecting only those drops which show only Chilomonas and bacteria to be present, rejecting all others. It is best to place these drops each on a small sterile coverslip, and then if the drop is found satisfactory merely drop the coverslip into the culture bowl. Now cover the cultures carefully with a glass cover or stack the finger bowls. In either case be sure that the covers are sterile and chemically clean. It is always wise to set up several cultures to insure against mishap. If successful the cultures will become more or less abundant in from four to six weeks. To subculture proceed as before except that it is not now necessary to wash the Amoebae, unless observation has shown that they have become contaminated with undesirable organisms.

The salt content of culture media While almost any uncontaminated freshwater will support growth it was found by Pace during a detailed study of the relation of salts to growth and reproduction that a more certain way of obtaining the proper salt concentration for abundant growth and reproduction is by using a synthetic spring water of a definite composition and concentration determined by experiment to be optimum for growth and reproduction. The following solutions are recommended:

Na_2SiO_3	15 mg	100 mg.
NaCl	12 mg	12 mg.
Na_2SO_4	6 mg.	6 mg.
$CaCl_2$	6.5 mg	6.5 mg.
$MgCl_2$	3.5 mg.	3.5 mg.
$FeCl_3$	4 mg.	4 mg.
Dist. Water 1000 cc.		1000 cc.

Sufficient HCI to give a pH of 7.0 to 6.8

We shall designate (1) as "dilute artificial spring water" and

(2) as "concentrated artificial spring water." The difference between the (1) and (2) solutions is in the concentration of sodium silicate. The first solution is recommended when attempting to culture Amoebae recently collected, since the concentration is nearer that of most natural freshwaters than is the second. However, when Amoebae have been cultured in a dilute medium they may be transferred to the more concentrated solution and more rapid growth and reproduction will be obtained. In fact the second solution is a solution in which optimum growth and reproduction was found to take place. A concentration between those of these two solutions in which the sodium silicate is 25 mg. per litre instead of 15 mg. or 100 mg. per litre, results in a much slower rate of reproduction. No serious difficulty will arise if the total salt concentration of (1) is a little less or that of (2) is a little greater, but the concentration of (1) must not be a little greater, or that (2) a little less than that indicated. Reproduction is much better when sodium silicate is present than when it is replaced by some other salt. Chalkley (1930) and Hahnert (1932) give salt mixtures which allow good growth and reproduction, but the "concentrated artificial spring water" just given has proven superior in our experience.

Hydrogen-ion concentration of cultures. Amoeba proteus will culture successfully at widely varying hydrogen-ion concentrations. Good cultures have been obtained at pH values anywhere between 4.0 and 8.5. A pH value of 6.6 to 6.8 is generally considered to be optimum. The optimum pH value, however, is probably dependent upon the salt content of the culture. Therefore for ordinary purposes it is' unnecessary to take any steps to control the hydrogen-ion concentration of the cultures. The great majority of cultures set up as described above will come to an equilibrium at a favourable hydrogen-ion concentration.

The maintenance of a constant hydrogen-ion concentration for special purposes is rather difficult. The addition of enough buffer salts to maintain a constant pH value brings the total salt concentration to a value too high for growth of the Amoebae. However, Hopkins and Johnson by a gradual addition of buffer salts were able to adapt the Amoebae to the increased salt concentration so that a constant pH value was maintained, and at the same time Amoebae grew and reproduced. This procedure should be as follows: To a new culture in which Amoebae have begun to grow and reproduce add each day about 0.5 cc. of Clark and Lubb's phosphate buffer of the desired pH value for each 100 cc. of culture fluid until a total of 5 cc. of the buffer has been added for each 100 cc. of culture fluid.

Hopkins used a feeding method for maintaining a constant hydrogen-ion concentration. Amoeba cultures when first set up as described above first become more acid in reaction and then gradually return to the alkaline side of neutrality. If, when in the return of a culture to alkalinity a given pH value is reached which is desired to be maintained as constant value, one adds a certain amount of fresh sterilized hay or wheat infusion daily, the acid tendency of the fresh infusion will oppose the alkaline tendency of the culture. By measuring the hydrogen-ion concentration daily and adding fresh infusion accordingly it is possible to maintain the concentration within a range of 0.2 pH units.

Temperature. As long as the temperature in the culture is below 25°C they remain in good condition. If the temperature is too low growth is retarded. It is necessary to freeze them before they are seriously injured. About 22°C. is perhaps optimum for growth. In summer when room temperature goes above 25°C. it is advisable to keep cultures in a cool basement room or, better, in a cold room where the temperature is maintained at the desired level.

Light. Direct sunlight is injurious. Growth is just as good in absolute darkness as in any light intensity.

Using methods described above the senior author has kept *Amoeba proteus* in continuous culture since 1923, a total of 12 years, and the Amoebae are still in excellent condition.

Culturing Amoeba dubia

Amoeba dubia is found in freshwater ponds and streams among aquatic plants such as Cabomba or Elodea, or in the debris on the bottom of such bodies of water, particularly among rotten leaves. Large individuals are often found in considerable numbers among Sphagnum.

Place small amounts of such material in finger bowls or large petri dishes, cover with spring water or with water from the source, and add 2 or 3 grains of uncooked rice, or an equal number of one-inch lengths of boiled timothy hay stalks. Do not place too much of the material in a single dish. This results in decay, and in the appearance of large numbers of bacteria which cause the death of any Amoebae that may be present.

Amoebae will appear in considerable numbers in successful cultures within a week or ten days. The decaying organic material is then removed and the Amoebae cultured by the following method. Make a hay infusion of 8 one-inch lengths of timothy hay stalks in 100 cc. of spring water, boil for 10 minutes and allow to stand for 24 hours. At the end of this time add large numbers of small Protozoa such as Colpidium and Chilomonas to the medium. Allow this to stand for two or three days before using. The Amoebae multiply rapidly on this medium so that the bottom of the culture dish is soon covered with them.

These cultures should be examined weekly. *Amoeba dubia* has been observed to ingest 50 to 100 Chilomonas within 24 hours. This results in the rapid disappearance of the food organisms from the culture. If the food organisms become few in number pipette off half of the culture medium and add an equal amount of fresh protozoan hay infusion. At the same time add 2 grains of uncooked rice, boiled wheat, or 4 one-inch lengths of boiled timothy hay stalks per 50 cc. of culture medium.

The Amoebae in such cultures divide rapidly up to 13 divisions per ten day period as judged from organisms kept in isolation. The cultures may be run successfully for as long as six months. They may be kept at room temperature even during the summer months though it is best to use refrigeration if the temperature reaches 90° F. or higher.

Subcultures are made by pipetting half of the material on the bottom of the old culture into a new dish and adding an equal amount of protozoan hay infusion and food material.

This culture method may be varied somewhat, but following facts should be kept well in mind. Large numbers of bacteria in a culture tend to cause the death of the Amoebae; therefore other Protozoa should be present in the medium. The use of large amounts of organic material such as boiled rice, wheat, or cracked wheat should be avoided. These ferment very readily, causing the death of most of the Amoebae. The depression period observed by many investigators is due to this cause. If the protozoan hay infusion with a small .amount of food is used about 90% of the cultures will be successful. This compares with only 50% of bacterial cultures in the author's experience. In the former case there is no depression period, in the latter a depression period of a month is not unusual.

Culturing Arcellae

Arcellae may be cultured in a hay infusion as follows:

Place 10 grams of clean timothy hay in a clean pyrex beaker containing 250 cc. of distilled water (pH 6.8). Heat to boiling point and allow to boil slowly for 5 minutes, then strain through two thicknesses of cheesecloth and store in quantities of about 3 cc. in small, sterile, hard glass test tubes. The tubes should then be plugged with cotton and placed in boiling water for 15 minutes. After 2 days they should be subjected again to boiling for the same period of time in order to kill any bacteria which may have escaped the first sterilization by being in the spore stage. The medium in these tubes constitutes the stock solution and will keep for months without deterioration provided evaporation does not take place.

In making up the culture medium take 1 part of the stock solution and add to it 9 parts of distilled water (pH 6.8), giving a 10% hay infusion. After the tube containing some of the stock solution has been opened and a part of its contents used the remainder should be discarded.

Using the medium prepared in the above manner Arcellae may be cultured in hollow-ground slides over a long period of time in a constant medium. The culture medium should be changed every day or two and the depression slides should be kept in a moist chamber placed in subdued light.

Methods of Culturing Testacea

Difflugia oblonga (D. pyriformis), small varieties; *D. lobostoma* and D. *constricta,* small varieties; and *Lesquereusia spiralis* may be cultured almost indefinitely by using a number of the green algae for food, such as Spirogyra, Zygnema, Mougeotia, and Oedogonium.

Culturing may be carried on in almost any type of container, though the shallow types usually give best results.

Spring, pond, or tap water may be used, with preference in the order named. The pH should be between 6 and 7.3. If small containers, depression slides, *etc.,* are used over long periods of time, the ;water must be changed frequently (every 2 to 3 days). Larger cultures in petri dishes may be run a week to 10 days.

Cultures should be kept out of direct sunlight and below 22° C, if possible. Small amounts of fine sand should be provided in cultures of D. *oblonga,* D. *constricta,* and D. *lobostoma* for shell construction.

Culture Methods for Marine Foraminifera

Although the Foraminifera are universally distributed in the sea and have been the subject of investigations for more than 200 years, comparatively little is known concerning their methods of reproduction, or of those physiological factors which limit the geographical and bathymetric distribution of these organisms. Confusion in the systematic designation of species is frequently due to their changing morphology which is the result of an alternation of generations, growth stages, a response to environmental conditions, or parasitism. In any attempt at a natural classification of these polymorphic forms, it is necessary to recognize their genetic relationship, and, where it is possible, their biological explanation should be determined. The solution of many of these problems can best be approached by means of laboratory cultures. Many species of the littoral zone may be maintained in cultures with a minimum of effort and without the use of running seawater. Therefore, this field of investigation offers a splendid opportunity for original work to anyone who has access to the sea and has acquired a reasonable amount to skill in microscopic technique.

In selecting a problem in this field, time is an important factor to take into consideration, due to the low rate of reproduction in this group as compared to other Protozoa. Growth in the majority of polythalamous species is a discontinuous process, because of an alternation of a vegetative phase with the addition of each newly formed chamber. In small species of Discorbis, where the test is composed of a continuous series of from 14 to 19 graduated chambers, from 3 to 12 hours are required for the addition of each new chamber. Under optimum conditions an individual will mature usually in from 19 to 23 days, and at that time produce from 30 to 40 young by multiple fission. In the larger species of Elphidium, in which the test consists of from 40 to 50 chambers, it is doubtful whether more than two generations occur annually. Since the life span determines the frequency with which one might expect to encounter individuals in a state of reproductive activity, it will be less difficult to obtain cytological evidence in support of a proposed life cycle in small quickly maturing species, and a more abundant supply of material will become available in a given time.

The following methods of collecting and maintaining these organisms in culture has proven satisfactory for species of Discorbis, Pyrgo, Triloculina, Bulimina, Patellina, Spirillina, and Robulus, and should be satisfactory for small species found within the limits of the intertidal zone.

An abundant supply of living Foraminifera may usually be obtained by washing seaweed or eel grass vigorously between the hands and allowing the organisms, sand grains, and other bits of debris to settle through a piece of bolting cloth into the bottom of a glass vessel. A convenient glass bucket for this purpose is made from a 10 x 14 inch battery jar provided with a rope handle and covered with canvas for protection. The bucket should be equipped with a tubular net 8 inches deep attached to a wooden hoop that will rest on the upper rim of the bucket. The vertical sides of the net should be made of unbleached muslin and the flat bottom of No. 00 bolting cloth.

After allowing the organisms about one minute to settle, the water should be decanted. Repeated washing by decantation will free the collection from silt and a considerable amount of organic debris that would decompose later. If several collections are to be made the material must be transferred to another container. A set of glass refrigerating dishes 6 inches in diameter and 2 inches deep that stack one on top of the other is convenient for this purpose. A carrying rack should be provided for the dishes and, where the collecting ground is some distance from the laboratory, it is advisable to control the temperature by packing with ice.

The rate of mortality is high in newly collected material, and we have found at La Jolla that certain species that will survive at room temperature for more than a year do not reproduce until the temperature is lowered to 18°C or less. Therefore, it is well to sort the material after the Foraminifera have become acclimatized to laboratory conditions.

Crowding of newly collected material should be avoided at all times. Not more than 5 cc. of the washings containing the Foraminifera should be placed in each of a number of 4 inch round-bottomed finger bowls filled with seawater, or about 20 cc. of this material may be added to a 10 inch crystallizing dish. The dishes should be covered to prevent excess evaporation and contamination.

The water should be changed twice a day for the first few days, and after that, once a day for a period of about two weeks. By that time many Foraminifera will have crawled up the sides of the dishes and, if a suitable substrate of diatoms has developed, several species should have become established and reproductive activity begun.

To establish persistent cultures of a single species, dishes should be prepared with a suitable substrate of diatoms before the isolation and transfer of the Foraminifera. Pure cultures of Nitzschia, Navicula, or similar diatoms may be used for this purpose, or substrate material may be used which has been taken from a dish in which the Foraminifera have become established and in which the diatom substrate is thin, uniform, and free from filamentous algae. If the latter method is employed and no new material is added to contaminate the culture, a single species of diatom will usually dominate the substrate in a short time.

The day after foraminifer has reproduced asexually, the young, which in some species number 200 or more, remain in the vicinity of the parent tests. In establishing subcultures, several thousand individuals may be transferred in a minimum of time by selecting these groups. If this method is employed the age of the organisms will be known and a maximum number of individuals of any stage of development will be available at a given time. A convenient mouth pipette for handling these organisms has been described.

After the cultures are established, it is advisable to change the water occasionally to compensate for evaporation and to replace nutrient material removed by the organisms. A strong stream of water directed against the sides of a dish by means of a glass syringe equipped with a large rubber bulb will remove accumulated debris and help maintain a thin clean substrate.

Before seawater is added to a culture, the water to be added should be filtered through a porcelain base Berkefeld or a sintered glass filter of suitable porosity. Growth of diatoms may be influenced by controlled illumination. When a subdued north light does not produce a suitable growth, a few drops of a saturated solution of potassium nitrate may be added to each culture or Allen and Nelson's modification of Miquel's solution may be used.

Temperature is a limiting factor in the distribution of species. In cultures of *Patellina corrugata* there is a difference of only 4° t0 5° C. between the optimum and the upper thermal limit at which this species can exist. Therefore, in any attempt to culture Foraminifera it is necessary to avoid temperatures above the mean that prevails in the sea during the summer months.

By employing the simple precautions herein described cultures of a number of species of Foraminifera have been maintained from one to three years, and on several occasions have been successfully transported over land a distance of more than 500 miles.

CULTURE OF AVIAN PLASMODIUM AND HAEMOPROTEUS

Since methods for the growth of avian parasites of the genera Plasmodium and Haemoproteus in the absence of living cells of their hosts have not been worked out, these forms must be maintained in one of their hosts. While successive generations of Plasmodium from man have been "cultured," this has only been accomplished by the daily addition of fresh erythrocytes, so that, strictly speaking, this amounts to culture *in vivo.* A method of cultivation has not been worked out which is successful in maintaining the strain over long periods of time. The invertebrate hosts of all of the species of avian Plasmodium whose life cycles have been worked out are culicine mosquitoes. Avian species of Haemoproteus are transmitted by parasitic flies belonging to the Hippoboscidae.

Strains of Avian Plasmodium

All strains of avian malaria of the genus Plasmodium may be maintained by inoculating blood from the infected into a normal bird. *P. relictum* and *P. cathemerium* may be easily transmitted from bird to bird by means of various culicine mosquitoes, the best vectors being *Culex pipiens* and C. *fatigans.* The mosquitoes must engorge on the infected bird at a time when gametocytes are present in the blood. Their bites are infectious for other birds after 8 to 14 days depending upon the species of the parasite and the temperature of the environment. *Plasmodium circumflexum* does not infect these common mosquitoes but it has been shown by

Reichenow to be transmitted by *Theobaldia annulata.* While several of the species of culicine mosquitoes become infected in small percentages when fed on birds with infection of *Plasmodium elongatum* and P. *rouxi,* no complete transmission has yet been effected by any of them.

Unless there is some special reason why mosquitoes are preferred as the means of transmission, all of these strains may best be maintained in canaries and passed when desired by blood inoculation. Infections produced by any of these species of parasites go through an acute stage which is followed by a latent period of infection of long duration. Spontaneous cure is rare. Therefore, in most cases, the strains may be maintained most easily as latent infections. New infections may be produced in normal birds by inoculating them with blood from the birds with latent infections. A few drops of blood are taken from a leg vein into physiological saline solution (0.85%) and injected by means of a syringe and inoculating needle into a normal bird. This may be done intraciently nutritive to permit the growth of contaminating micro-organisms. Justification for this may be seen in the fact that molds were never observed to develop during the experiments.

All experiments were carried out in a constant temperature room at 72°, with north light.

The containers for the culture of Colpoda are watch glasses of about 2 mm. diameter enclosed in petri dishes of suitable size containing in the bottom a small amount of water and so used as moist chambers.

Separation of the Protozoa from Other Micro-Organisms

Migration through Pipettes. The technique developed by Glaser and Coria, and used by them for the separation of certain Protozoa from their contaminating bacteria, depends essentially on the exhibition by the Protozoa of a geotropic response which causes them to swim away from the bacteria through a column of sterile liquid. Two methods for bringing this about are available. In the first, sterile pipettes are used, at least 14 inches long with a 1/4 inch bore, a tapering point, and a cotton plug at the large end. For negatively geotropic Protozoa such a pipette, by suction through a rubber tube attached to the large end, is filled with sterile tap water to within 2 inches of the top. Then about 2 cc. of a heavy culture of the contaminated Protozoa is carefully sucked up into the pipette, so as to form a layer beneath the sterile water. The tapering end of the pipette is then sealed by heat, care being taken not to permit the formation of air bubbles. The pipette, sealed end down, is set upright in a test tube rack. In from 5 to 30 minutes, negatively geotropic Protozoa will be present at the top of the column of liquid. Even motile bacteria, by their own efforts, can not reach the top in so short a time. Usually, however, one such washing does not free the Protozoa from adherent or ingested bacteria, and in most cases it is necessary to repeat the procedure once or twice, each time sucking up beneath the fresh column of sterile water in a fresh pipette about 2 inches of fluid from the top of the previous pipette. In some cases it is advisable to leave the first or second pipette for 18 to 24 hours to give the surface migrants a chance to evacuate the remains of ingested micro-organisms and to multiply to some extent. This procedure is followed by another rapid washing. Finally a drop of the surface fluid is inoculated into a tube of sterile medium. In this manner the ciliates *Trichoda pura,* three strains of the thermal ciliate *Saprophilus oviformis* from Hot Springs, Virginia, a large undetermined ciliate, *Paramecium caudatum* and *P. multimicronucleatum,* and the flagellates *Chilomonas paramecium, Parapolytoma satura,* and an undetermined monad from the intestine of the fly, *Lucilia caesar,* were all freed of bacteria, as shown by consistently negative findings in stained films and aerobic and anaerobic cultures on routine laboratory media at both room and incubator temperatures. The upward migration here involved is a genuine geotropic reaction, as it occurs in pipettes held in the dark and in those sealed without any air space at the top.

This same pipette method may be used for positively geotropic Protozoa. The pipette is nearly filled as before with sterile water, the tip is sealed and then a small amount of the contaminated culture is layered on top of the sterile liquid. After a suitable time the end of the pipette is cut off and one or two drops either inoculated into culture medium or placed at the top of another water column for a second washing. In this way an undetermined free-living monad was freed of bacteria after two washings, each consuming about 30 minutes. With either negatively or positively geotropic organisms the addition of killed yeast cells to that part of the water column toward which the Protozoa were to migrate, accelerated their migration.

Migration through V-Tubes. In the second method, V-shaped

tubes are used, one arm of the "*V*" being 12 cm. long with an inside diameter of 28 mm., the other 9 cm. long with an inside diameter of 8 mm. The tube, after sterilization, is filled with 15 cc. of melted semi-solid medium. When this has set to a soft gel the contaminated Protozoa are introduced by means of a long fine capillary through the small arm into the bottom of the large one. The tube is permitted to stand for a length of time dependent of the Protozoa concerned, and then samples are taken from the surface of the medium in the large arm. One such treatment frequently suffices. A modification of this technique was used to separate *Spirillum undulans* from other bacteria. A loopful of contaminated culture of the Spirillum was placed on the surface of sterile tap water in the large arm of a V-tube. Five-tenths of a cc. amounts, withdrawn one hour later from the surface of the small arm, gave pure cultures of the Spirillum upon inoculation into suitable media.

Culture Media for Certain Bacteria-Free Protozoa

"*Basic Medium.*" Among the free-living Protozoa freed of bacteria by any of the above methods some, such as *Trichoda pura,* were able to grow well in simple media such as peptone water. Even such Protozoa, however, grow better in the so-called "basic medium" of Glaser and Coria. This has been recently modified to give more nearly uniform results and is now prepared in the following way: Stock solution *A* consists of 50 cc. of horse serum in 1000 cc. of well water (or distilled water). This is autoclaved for 30 minutes at 15 lbs. and has a pH of 7.0 without adjustment. Stock solution *B* consists of 50 gms. of timothy hay in 1000 cc. of water. The mixture is infused over night in the refrigerator, filtered through cotton, and the reaction adjusted to pH 7.2-7.4. The stock solutions *A* and *B* are stored separately in a refrigerator and when needed are combined with well water in these proportions:

Solution A	500 cc.
Solution B	250 cc.
Well water	250 cc.

The medium is adjusted to pH 7.2-7.4, tubed in 8 or 10 cc. amounts and autoclaved. This final medium was used only for Protozoa free of other micro-organisms. When heavy initial cultures of unpurified Protozoa are desired 2 cc. of the final medium is added to 20 cc. of the original Protozoa-containing water. To make a solid medium, 1.5 to 2% agar is added. For a semi-solid medium similar to Noguchis Leptospira medium, 100 cc. of the melted solid medium is diluted with 900 cc. of warm water.

Raw Potato. Some Protozoa, which grew well in basic medium when contaminated with bacteria, did not grow at all in this medium when free of other micro-organisms. Thus a flagellate from the intestine of *Lucilia caesar* would not grow in basic medium except in the presence of living bacteria. The bacteria-free flagellate would not develop on autoclaved potato, but delicate growths of it were obtained in tubes of raw potato under tap water. Such tubes are prepared as follows: Raw potatoes are scrubbed with hot water and partly dried by heat; a part of the surface is washed thoroughly with 70% alcohol and flamed until charred. Cylinders are then cut out with a sterile No. 5 cork borer and put in sterile petri dishes where they are cut into pieces 1/2 inch long. Each piece is placed in a tube of sterile water. Such tubes are held at room temperature for some time before use, and any showing contamination are discarded.

Partial Purification of Cultures of *Balantidium coli*

It has long been realized that bacteria-free cultures of the intestinal protozoan parasites of vertebrates would be extremely useful for the study of the biology and pathologenicity of these organisms. Yet up to the present time no such protozoan has been grown in culture free of bacteria and in most cases it has been impossible even to free the Protozoa of the numerous bacteria which naturally accompany them. Cleveland, however, succeeded in freeing the coprozoic organism, *Tritrichomonas fecalis* of man, from bacteria and was able to grow it on heat-killed bacteria. Glaser and Coria have recently applied one of their migration techniques to *Balantidium coli* from swine and have been able to free the organism of all bacteria except *Bacillus coli.* Largely as a result of this partial purification, they have been able to maintain this parasitic ciliate in better condition and for a much longer time without subculturing than has heretofore been possible.

They used a liquid medium consisting of 9 cc. of sterile Ringer's solution, 0.5 cc. of sterile horse serum, and a sprinkle of rice starch in each tube. Much better result were obtained with a semi-solid medium. This is prepared by adding 25 cc. of 2% standard nutrient agar, 1 gram of sterile rice starch and 12.5 cc. of sterile horse serum to 250 cc. of sterile Ringer's solution warmed to

50°C. The pH of the mixture is adjusted to 7.2 to 7.4 and 15 cc. amounts are transferred aseptically to sterile tubes.

"V" tubes containing 15 cc. of semi-solid medium were inoculated (after being warmed to 37°C.) on the surface of one arm with material scraped from crypts in the mucosa near the iliocecal valve of a pig. The Balantidia migrated downwards within 24 to 48 hours at 37°C. and could then be recovered in a sterile pipette inserted through the uninoculated arm of the "V" tube. The semi-solid nature of the medium checked the spread of bacteria, which in a liquid medium would rapidly grow over both arms of the "V" tube. The Balantidia were able to push their way through the gelatinous semi-solid medium and it was found that in such a medium a much higher percentage of positive initial cultures could be obtained than in a liquid medium. Balantidia removed after the final migration were again placed on the surface of one arm of a fresh "V" tube and again permitted to migrate for 24 to 48 hours. The procedure was repeated five times, after which the ciliates were further cultured in ordinary tubes of semi-solid medium. Strains which would not originally grow in the liquid medium could be adapted to this after varying lengths of culture in the semi-solid medium.

These methods failed to free Balantidium of *Bacillus coli* but they did yield cultures which might be regularly transplanted at 8-day intervals and one strain (in liquid medium) which was transferred every 20 days thrived for over 2½ years.

PARAMECIUM

Paramecium is a form which is easily reared. From the bottom of a permanent pond the foul-smelling debris is taken and kept in a bowl barely covered with water and at a temperature of approximately 73° F. As the debris fouls, the Paramecium become abundant. From time to time (about once a week) a half-inch cube of fish is added to the bowl to maintain a supply of food. Such a culture as this will tarry on for months. It is advisable to select a large race and to rear in separate containers. Should some small forms be present in the new bowls they are usually unsuccessful in the competition with the large ones.

A Culture Medium for *Paramecium*

This medium has proven very satisfactory for culturing

various species of Paramecium in pure line cultures. The main result of the use of the medium is that the organisms do not exhibit a lowering of their normal metabolism after continuous culturing.

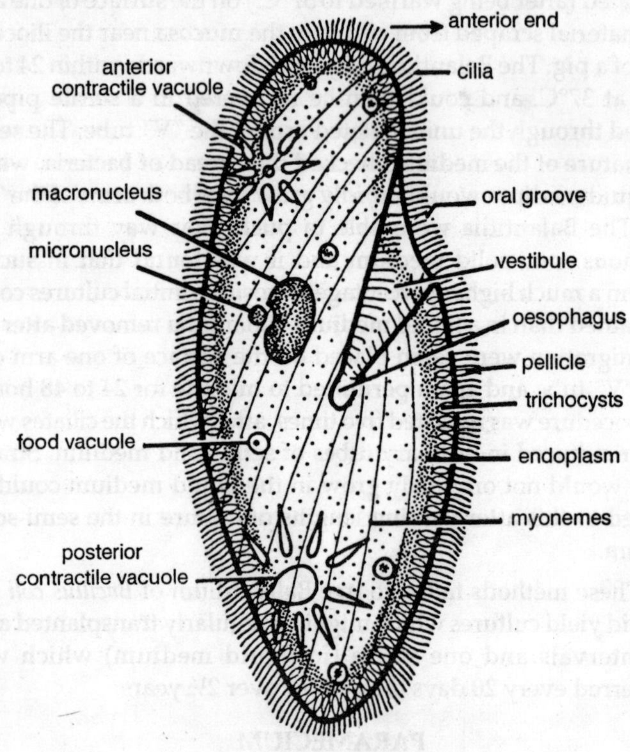

Fig. 1.1 *Paramecium*

The basic part of the medium is the usual hay infusion of 10 grams of chopped timothy hay boiled for 15 minutes in a litre of well water. This infusion is filtered and sterilized in the Arnold sterilizer at 100° C. one hour a day for 3 days. It is diluted with 9 volumes of sterile well water just before using. Two portions of this infusion are placed in sterile litre flasks with sterile cotton stoppers. One liter is inoculated with *Bacillus subtilus* and the second with B. *coli communis*. A third portion of the medium is made up as follows; Approximately 30 grains of wheat are boiled in a small amount of water for 10 minutes. The wheat grains only are then placed in a third liter flask of sterile well water. The three portions are incubated at 37° C. for 24 hours, and then combined in one large sterile flask. The medium is now ready for use. The culture may be used in almost any size of container, but that used has been the 300 cc. Erlenmeyer flask. These flasks are fitted with cotton stoppers and sterilized. Each flask is filled about 2/3 full of the medium, and different species are transferred to the cultures with sterile pipettes.

The original basic infusion may be made up, sterilized, and stored in a refrigerator until ready for use. Likewise, the medium, made of the three portions, may be stored in a refrigerator for later use.

Sterile precautions are maintained throughout the procedure, but after Paramecium has been transplanted, such strict precautions are no longer necessary .The essential part of the process is to provide a medium rich with a suitable food in which Paramecium will continue to grow normally. With these sterile precaution, other ciliates and flagellates are eliminated. A single organism placed in such a medium will produce a flourishing culture in 7 to 10 days. One should transplant every 2 to 3 weeks. *Paramecium multimicronucleatum, P. bursaria,* and *P. aurelia* have thrived in this medium.

Cultivation of *Paramecium aurelia* and *P. multimicronucleatum*

Of the many media tried, the most successful is an infusion of 1.5 gms. of desiccated (but not burned), powdered lettuce boiled for 3 minutes in a liter of double distilled water. This infusion is filtered while hot, dispensed into pyrex test tubes or flasks, containing an excess of pure $CaCO_3$ (which adjusts the pH to about 7.2), plugged with cotton, and autoclaved. When ready for use, it is filtered to eliminate the $CaCO_3$ and is inoculated with the bacterium, *Flavobacterium brunneum,* grown on beef-agar slants, and the alga *Stichococcus bacillaris,* grown on 0.05% Benecke's agar slants. The bacterial slants may be used when 1 to 5 days old; the algal slants when 18 to 24 days old. The quantities inoculated into the culture fluid are one 1-mm, loop level full of bacteria and three 2-mm. loops of algae to 20 cc. culture fluid.

Isolation cultures may be carried on depression slides by using two drops of this fluid per depression. Such cultures must be renewed daily by transferring one Paramecium to freshly inoculated fluid on a fresh slide. The optimum temperature is 27°-28° C.

Mass cultures may be carried in cotton-stoppered flasks by 'using the same fluid. Such cultures must be made frequently to keep the Paramecia in good condition, but the organisms will live for three months or more in a greatly depressed condition without renewal of fluid,.

Mass-culturing

Variety in composition is one of the outstanding characteristics of infusions in which Paramecia are to be mass-cultured. There appears to be no single method of successful culturing In general, *small* quantities of materials, usually organic, must be introduced into water to induce bacterial multiplication. These bacteria are the chief food supply of the Paramecia, at least in the earlier stages of the culture. Liebig's beef extract, mangle beet water, lettuce leaves, wheat bananas, timothy hay, Horlick's malted milk, gelatin or curd placed under earth, meat, pond lily leaves red cabbage leaves, and soil are some of the materials which are reported to have been used. Such diverse organic materials may be boiled in water, to they may be allowed to macerate. When the infusion is ready, it may be autoclaved, especially if portions are to preserved for future use; or it may be used without sterilization.

Slight variations of the above techniques require the introduction of algae, yeast, or bacteria into infusions such as the above. Certain investigators adjust osmotic pressures, pH, or other ionic concentrations by adding salts or other compounds. Such techniques are not essential to the production of excellent mass-cultures of Paramecia, although they will enhance the uniformity of the medium.

The present discussion is limited to cultures of *P. multimicronucleatum* grown in infusions prepared by boiling hay or hay-flour combinations in water. The following formula has be employed frequently and successfully:

1 gram hay
0.1 grams white flour
700 cc. distilled water

Stir the flour through the hay. Bring the water to a boil. Add the hay-flour mixture. Boil 10 minutes. Cool. Add distilled water to replace that evaporated. Seed with 200 Paramecia on the second day.

The above formula may be considerably altered, especially in routine work, without materially interfering with the production of a satisfactory population. Permissible variations include the use of hay up to 4 grams if flour be omitted (Jones, 1930); the use of tap or pond water if non-toxic; and variation in the date of seeding and the number of seed Paramecia introduced.

Additions of hay or flour in excess of the amounts stated will produce a hydrogen-ion concentration which will destroy the Paramecia on the fourth or fifth day (pH 4.83 or less). The "seeding" must, under such circumstances be delayed until the pH exceeds 5.0. In the presence of such excesses of food, certain other unidentified split products which result may prove to be toxic, even if the pH be satisfactory .Introduction of excesses of food material must therefore be avoided, either when originally preparing the infusion, or later, if feeding techniques be employed.

Cultures should not be covered if populations of maximum concentration are desired. Non-evaporating cultures made as described above will produce populations of approximately 300 Paramecia per cc. of solution as a maximum, whereas evaporating cultures may eventually yield from 2000 to 4000 per cc.

Maintaining and Rehabilitating Mass-cultures

Mass-cultures, especially of the non-evaporating type (in which the cover is placed on the jar, but not screwed down), may be revived by feeding when they approach the point in the cycle of culture conditions at which the Paramecia would normally disappear. When a population of Paramecia is present in a culture it may frequently be increased by the same means.

Any simple method of introducing such substances is usually satisfactory if the materials are in a finely divided state or a fluid condition. Flour is introduced by sprinkling approximately 1/2 gram upon the surface of 700 cc. of infusion in as unlumped a condition as possible, and stirring. A solution of egg may be prepared by breaking an egg into 100 cc. of distilled water in a flask, and shaking with glass beads.

Other investigators report temporary revival of declining cultures secured by introducing sugar, bread, or fresh hay infusion and the culture has been maintained for a year, in one instance by occasionally introducing a few grains of wheat and a few pieces of timothy hay.

In the controls, which developed and declined in the usual mass-culture fashion, Paramecia could be found for a period of approximately six months. Dense populations were not present after the initial three months *The fed culture maintained Paramecia for more than twenty months.* It produced a dense population when fed when fourteen months old, although the Paramecium population had dropped almost to zero while the culture had gone unfed over a summer's vacation which had just preceded The Paramecia eventually disappeared for no known reason, other than that the culture had not been fed for some time.

To further test the potential longevity of Paramecium cultures, fourteen cultures of a capacity of one gallon were prepared. These received infusion of approximately the same composition, and seed in the same ration per volume as described above.

It became necessary to feed flour to all of these gallon cultures before they were six months old, because, in every culture the Paramecium population had declined almost to extinction. Six cultures now survive. These cultures have been fed repeatedly, usually only when the Paramecium populations were quite low, whereupon they have characteristically produced larger populations. Flour has been most frequently fed, but milk, glucose, albumin, and ethyl alcohol have been used at times. An epidemic of mold which destroyed the Paramecia appeared in practically every culture to which ethyl alcohol was fed.

At the present time, the six of the series which now retain populations of Paramecia are 30 *months of age*. To this group I have recently fed flour at approximately weekly intervals to determine whether dense populations can still be produced. Four of the cultures now contain an average concentration of 300 to 400 per cc. Such concentrations are high for large cultures which are not evaporating. To the best of my knowledge, none of these cultures has at any time contained a higher concentration.

Reductions in length and number of Paramecia which are characteristics of the declining culture. The corresponding increases which resulted when egg was fed are likewise indicated.

Securing Concentrations

These methods have been employed for concentrating either rotifers or paramecia. There is reason to believe that a considerable variety of protozoans and small metazoans will respond in like fashion.

Concentration not needed for two days. Distribute cultures which contain Paramecia long containers having the approximate dimensions of quart fruit jars, filling each container half full. Add to each enough freshly prepared, cooled infusion, made according to the formula given above (variation is permissible), to fill it completely. Do not cover. Within 40 to 60 hours the Paramecia will congregate on the sides of the container at or immediately below the surface. Remove the concentration with a pipette having a finely drawn tip and a bulb.

Concentrations for immediate use. Such concentrations may usually be picked up directly from the bottom of an older culture, if a longer pipette is employed. The Paramecia secured by such methods are usually smaller than those concentrated by the first method.

Debris-free concentrations. Paramecia may be freed from the culture debris by introducing into concentration tubes the animals and infusion taken by either of the preceding methods. For this purpose I have employed glass tubes which were 30 cm. long and which had an internal diameter of 8 mm. The organisms, following introduction, will settle to the bottom, after which they will systematically migrate to the surface, from which position they may be removed with a pipette.

Culturing *Paramecium caudatum* in Oat Straw Infusion

An excellent medium for culturing *Paramecium caudatum* for a rapid population growth is an oat straw infusion. After having experimented with various types and concentrations of culture media made from timothy hay, oat straw, barley straw, oak leaves, elm leaves, clover, alfalfa, *etc.*, oat straw was found to be the most favourable medium for rapid growth of my strain of Paramecium.

Cut 15 grams of oat straw into short lengths (1-3 cm. long) and place in a quart glass container. Pour 900 cc. of boiling distilled water over the straw. Plug the container with cotton and allow the mixture to cool. Adjust the pH to 7.8 with NaOH. The colorimetric method is sufficiently accurate. Keep the mixture at approximately 25° C. temperature for 48 hours. Shake the culture medium until it is thoroughly mixed; again adjust the pH to 7.8 and the infusion is ready to use. Add approximately 250 Paramecia and in a few days a mass-culture should have developed.

For best results make new culture medium every 48 hours and make new inoculations as often, because usually after 72 hours have elapsed the culture medium begins to deteriorate and is not at its best for optimum growth.

By following the above procedure a number of times a colony of rapidly dividing *Paramecium caudatum* can be developed, each animal dividing at an average rate of once every 8 hours. This is considered optimum growth under these conditions.

For culturing animals for classroom use put two or three dozen grains of oats in 1,000 cc. of water and allow the mixture to stand for three days before inoculating with Paramecia. Within a week a mass-culture of the animals usually develops. It is best to keep the culture covered when it is not in use.

A Culture Method for *Oxytricha fallax*

Oxytricha may be grown in 0.1% lettuce infusion. The lettuce for this infusion is obtained by drying lettuce leaves in an oven and pulverizing. The proper weight of lettuce is boiled in 0.005 M. KH_2PO_4 solution for 3 minutes, after which the infusion is titrated to a pH of 7.0 with NaOH. The particles of lettuce may be left in or removed. The culture lasts longer if the particles remain. Usually 15 cc. of such an infusion was seeded with 20 to 100 Paramecia of Oxytrichae. It is quite satisfactory for rough work to use tap water instead of the buffer solution.

Culture of *Nyctotherus ovalis* and *Endamoeba blattae*

Nyctotherus ovalis from the hindgut of the cockroach, *Blattella germanica,* may easily be cultured in a modified Smith and Barret medium. This medium was used by the discoverers for *Endamoeba (Entamoeba) thomsoni,* and according to Lucas, it is suitable for the cultivation of neither *Endamoeba blattae* nor *N. ovalis.* The medium used by Smith and Barret consists of 19 parts of 0.5% NaCl to 1 part of inactivated human blood serum. By substituting non-inactivated rabbit serum for the human serum a medium is produced in which *N. ovalis* lives and multiplies freely. Dividing forms are common, and occasionally precystic and cystic forms are met with. Three cultures have been maintained for 40 days and at the last examination the organisms were as normal in appearance as those found in their native habitat. Subculturing is done at weekly intervals, and the cultures are maintained at room temperature.

The cultivation of E. *blattae* has been less successful than that of *N. ovalis.* Two cultures out of twelve attempts were maintained for 29 days. At the end of this time the organisms were few in number entirely normal in appearance and movement. One 2-and one 8-nucleate form were seen, the latter with nuclei of different sizes and evidently precystic. The next examination was negative. This gradual dwindling in number does not necessarily indicate an unfavourable environment, but rather that division is not frequent enough to permit weekly subculturing without gradually diminishing the number of organisms to the point of extinction. Longer intervals between subcultures result in an over-growth of bacteria and the small flagellate *Monocercomouas orthopterorum.*

A Method for Culturing *Bursaria truncatella*

Add 1 gram of timothy hay, 1 gram of rye, and 5 grams of fresh cabbage to 600 cc. of spring water. Boil slowly for 5 minutes. Let stand uncovered for 2 days to allow development of bacteria, then remove the cabbage, add 500 cc. spring water, transfer 250 cc. of the solution with corresponding amounts of hay and rye to each of several 1/2 litre jars and inoculate with Paramecium, Colpidium, and Chilomonas. After 2 or 3 days, *i.e.,* when these organisms have become abundant, inoculate with Bursaria. Cover the jars and keep them in indirect sunlight at room temperature.

If a film of gummy substance has developed on the surface of the infusion, break it. If no gummy substance is present, add some from an old culture.

A culture thus prepared reaches a flourishing condition (10 ± individuals per cc.) in 2 or 3 days, and continued in this condition for 3 or 4 days. If the infusion is more dilute, the cultures flourish longer, but the Bursaria does not become so abundant.

CULTURE OF EUGLENA

Euglena may be cultured very readily for laboratory use in filtered rain water 100 cc.) to which is added 1 cc. of one day old pasturized milk from which the cream has been removed completely. Keep the culture at ordinary room temperatures of from

600 to 68° or 70°F. Place the culture near a window; diffused light is quite satisfactory. From three to six weeks are needed for the development of the culture. This time will depend on the density of the population desired. Ordinary half-pint milk bottles make satisfactory containers, and absorbent cotton stoppers, which will admit air, may be used to keep out dust and spores of other forms of life. Good results have been obtained in culturing other species of Euglena in the same manner.

A pure culture of Euglena may be obtained by selecting under the microscope with a fine pipette individual specimens of the desired species. A clean culture is very valuable and, once it is obtained, it may be perpetuated indefinitely if care is taken to prevent contamination. Obviously, only sterilized pipettes should be used in removing specimens from the culture for class use.

A Culture Medium for Free-living Flagellates

The following culture method, which has been tried out for two years, may be of use to other laboratories. Whole wheat is weighed into 5 gm. lots, which are then put into large test tubes and 25 *cc.* of tap water added. These are then plugged with cotton, capped with lead foil, and autoclaved at 15 pounds' pressure for 2 hours, which very thoroughly macerates the wheat. Tap water is again added up to 50 cc., and desired percentages of this fluid are used after shaking. After opening a tube it is necessary to sterilize again in an Arnold sterilizer, as bacterial growth is vigorous in the mixture. However, a tube may be used day after day, if sterilized daily.

Various percentages of this mixture afford a very good medium for many Protozoa. Bacterial feeders such as Chilodon, Paramecium, Oicomonas, and others thrive on it. Ochromonas, Chilomonas, and several of the smaller Euglenas *(E. gracilis, E. quartana,* and *E. mutabilis)* have been grown in abundance in various dilutions. There are several species of Amoeba which likewise occur or are capable of being cultured in large numbers. It has proved best, however, for Entosiphon and Peranema. Both of these forms are easily grown in quantities sufficient for classroom use; isolation cultures of the former have been carried over a year on this medium. In general it seems much better than cracked boiled wheat, which is often used.

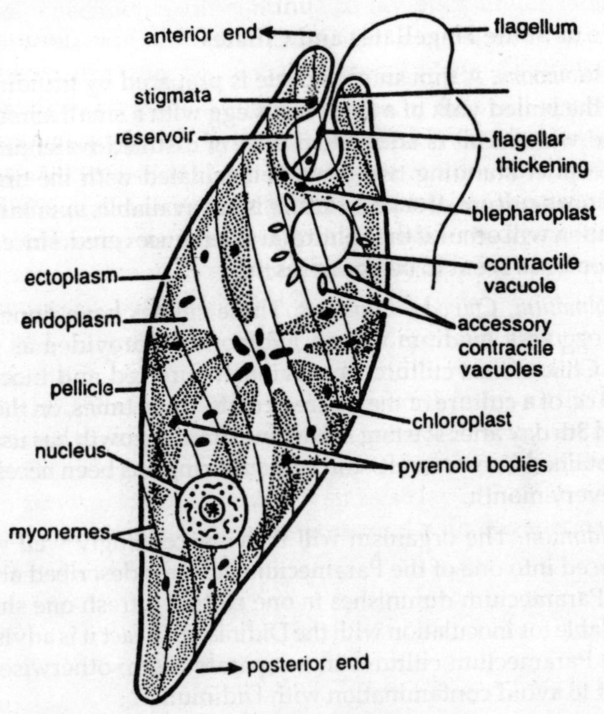

Fig. 1.2 *Euglena.*

Another method for *Euglena.* To 100 cc. of a modified Klebs' solution (Solution B) in a white glass battery jar add 40 rice grains (boiled 5-10 minutes) and 900 cc. of distilled water. The foregoing medium is allowed to stand for about five days. The jar is then placed in indirect sunlight (the direct rays of the sun should not strike this culture for more than an hour a day), and inoculated with Euglenae three times (10 cc. of a dense Euglena culture) at three day intervals. If an old Euglena culture is available the organisms may be found encysted on the sides of the vessel and it is of great advantage to inoculate the cysts along with the free Euglenae. Starting ten days after the initial inoculation, growth may be accelerated by adding (three times, at weekly intervals), 25 cc. of Solution B and 10 mg. of the tryptophane powder. The further addition of 5 grains of boiled rice each month will serve thereafter to maintain the culture. Large ciliates and rotifers are detrimental.

Another technique, applicable to several Protozoa involves the use of an egg yolk-distilled water medium.

Culture of Some Flagellates and Ciliates

Chilomonas. A thin smooth paste is prepared by grinding 0.5 gm. of the boiled yolk of a fresh hen's egg with a small amount of distilled water. This is added to 500 cc. of distilled water and the mixture, after standing two days, is inoculated with the original Chilomonas culture. If such a culture is not available, spontaneous inoculation will occur if the culture jar is left uncovered, since cysts of Chilomonas seem to be omnipresent.

Colpidium, Colpoda, Euplostes. These ciliates have done well on the egg yolk medium when Chilomonas is provided as prey. Start a Chilomonas culture as previously directed and inoculate with 10 cc. of a culture of the desired ciliate, three times, on the 4th, 6th, and 8th day after starting. Dense maximum growth has usually been obtained in two weeks and sub-culturing has been necessary about every month.

Didinium. The organism will thrive exceedingly well when introduced into one of the Paramecium cultures described above. As the Paramecium diminishes in one culture a fresh one should be available for inoculation with the Didinium. In fact it is advisable to keep Paramecium cultures in a separate room; otherwise it is difficult to avoid contamination with Didinium.

Vorticella. A modification of the method found useful for Paramecium is necessary here. The medium of 1/2 gram of mashed hard-boiled egg yolk in 750 cc. of distilled water is permitted to stand for two days; it is then filtered through cotton, 100 cc. of the filtrate are added to a finger bowl and the Vorticellae are introduced; great numbers of the animals will be found clinging to the glass surface within two weeks. It is advisable to subculture every three weeks.

Stentor. This organism may be cultured by two methods; namely, that found useful for Amoeba, and that described for Vorticella, except that certain modifications must be made here. The rice-agar with 50 cc. of Solution A is permitted to stand for two days in a finger bowl; at this point Chilomonas is added together with as many Stentors as possible. Or, the medium *of* ½ gram of mashed hard-boiled egg yolk in 750 cc. of distilled water is permitted to stand for three days, filtered, and inoculated with Chilomonas and the Stentors. (About 5 cc'. of a heavy Chilomonas culture are necessary for the inoculation in both cases.) it is desirable to subculture every month.

Stylonychia and *Oxytricha*. For these hypotrichs and certain others the presence of Chilomonas seems advantageous. Before introducing the desired ciliate, allow some 30 cc. of Solution A in a rice-agar bowl, inoculated with 10 cc of a Chilomonas culture to stand about four days. Swarming cultures have usually been obtained within two weeks.

CULTURE OF TRYPANOSOMA

It is not always feasible, on short notice, to obtain living trypanosomes. Laboratory cultures should be established and maintained. This can be done easily by using the following method which we have found both simple and reliable.

Wild rats are caught by setting a number of spring-traps near their holes or runways. The captives are killed and opened. A small amount of blood is taken from the heart, using clean pipettes, and this is dropped into cotton-stoppered vials half full of sterile 0.75% saline solution. The vials are taken to the laboratory, a drop of the fluid placed on a slide, covered, and examined. Under low power (16 mm. objective) the parasites cannot be seen, but if they are present their movements cause violent agitation of the corpuscles, thus enabling a tentative diagnosis to be made. They are readily seen under high power (4 mm. objective) if the light is carefully regulated. About 5 to 10% of the wild rats carry infections.

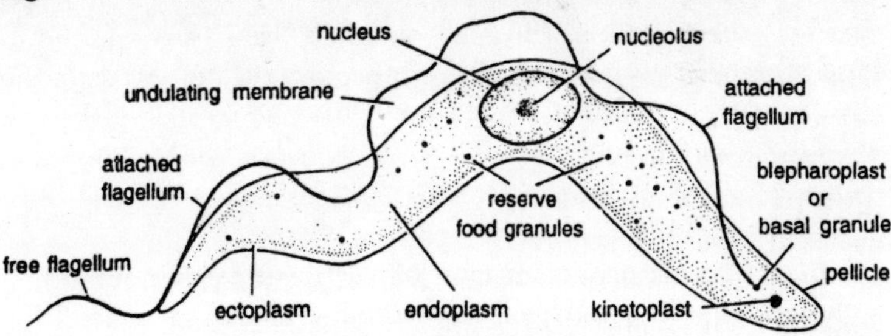

Fig.1.3 *Trypanosoma*

To establish the strain in laboratory animals the fluid from positive vials is drawn into a clean hypodermic syringe, and intraperitoneal inoculations are made into partly grown laboratory rats, about 0.5 cc. being given each one. Occasionally an animal will be found to be immune. This fact necessitates the infection of several so as not to lose the infection. About seven days later a test is made. The tip of the tail is cut off and a drop of blood allowed to mix with a small drop of saline solution on a slide. It is then covered and examined microscopically.

By the end of the second week the trypanosomes have usually reached their maximum number and have begun to decrease. By the fourth or fifth week (occasionally somewhat longer) they may have disappeared entirely from the blood. Before this time it is necessary to inoculate new, non-immune animals. An infected rat is etherized; the needle-of a hypodermic syringe partly filled with saline solution is inserted into the heart; and some blood is drawn out. This is injected directly into the peritoneal cavity of the new animals. By repeating the transfer of infective serum at the proper times the strain may be maintained for indefinite periods.

Occasionally it is desired to keep living trypanosomes for a few days in the serum-saline mixture. This may be accomplished by placing the container in a refrigerator (about 18°C.) where they will live for a week or longer.

2

Culture of Sponges

Sponges *(Stylotella heliophila)* are placed in outside aquaria, concrete or wooden tubs, covered with glass and not in direct sunlight. The sponges should be clean, raised from the bottom on bricks; half a dozen to an aquarium 60 cm. in diameter and 30 cm. deep. The aquarium (tub) is emptied, filled, and flushed for some minutes three times in every twenty-four hours. Reduction begins in a day or two. In the course of two or three weeks gradual death of the tissues coupled with reduction leads to the formation of many small living masses of varying shape lodged on and through the skeletal network of the sponge. In the most striking cases these masses are numerous, more or less spheroidal and small, 1 to 1 ½ mm. in diameter. Such a dead and macerated sponge body with its contained nodules of brightly coloured living tissue suggests a spongilla full of gemmules. The histological structure of nodules varies with their age, but is very simple, although many details in the process of reduction are unknown.

The reduction bodies have regenerative power. If enclosed in bolting cloth bags and hung in a live box they quickly transform into sponges. Probably with a very excellent water supply the transformation could be induced in the laboratory.

Similar bodies have been produced in the Calcarea and in freshwater sponges.

Growth of Sponges from Dissociated Cells

A sponge *(Microciona prolifera)* is cut up into small bits about 3 mm. in diameter, which are allowed to fall on a piece of fine bolting cloth supported on the edge of a slender dish and semi-immersed. The cloth is then folded like a bag around the bits of sponge, is partially immersed in a small dish of filtered seawater, and while it is closed with the fingers of one hand is repeatedly squeezed between the arms of a small forceps. The pressure and the elastic recoil of the skeleton break up the tissue into its constituent cells and these pass out through the pores of the cloth into the surrounding water. The cells fall to the bottom and may be sown with a pipette on any desirable substratum (slide, cover glass, or oyster shell) immersed in a culture dish of seawater. Or the cells, as pressed out, may be allowed to fall at once on the definitive substratum. The cells attach in the course of an hour or so and the slide or other body may be removed to a fresh dish of water or to a running water aquarium, where it should be raised well above the bottom and protected from the force of the current. Attachment will usually take place by means of coarse reticula which remain permanently attached. Reticular pieces if partially freed from the slide will curl up and form balls, the size of which is under control. Such balls may be transferred to slides in other dishes or in running water aquaria. A convenient vessel to use is a porcelain-lined bucket, in which the slides rest on inverted bottles and are so brought near the surface of the water, the current entering at the bottom. By changing the water two or three times a day, such an arrangement serves in place of 'a running water aquarium. The balls attach and sponges of desired size may be obtained. The attached reticula or balls metamorphose and in the course of a few days will have transformed into incrusting sponges with functional canal systems. Such cultures are easily kept for long periods of time in small wire guaze cages hung in live boxes. Lobular outgrowths and even embryos have developed in sponges treated in this way.

Growth of Sponges from Ciliated Larvae

Mycale (Esperella) fibrexilis has embryos during July-August at Woods Hole, Mass. Larvae are liberated, sometimes at once, on placing the sponge in an ordinary 2 gallon glass aquarium jar. Larvae are picked out with a pipette and transferred to culture dishes where they may be kept by changing the water several times a day. They attach in a day or two. They may be made to attach to cover glasses or if they are to be used as section material it is convenient to coat the dish with a thin layer of paraffin and let them attach to this. Little pieces of paraffin with the attached and metamorphosing larvae may then be cut out, fixed (paraffin and sponge), and hardened, the sponge often detaching itself from the paraffin.

Other halichondrine sponges breed during the summer (July-August). Larvae may be reared in the same way or may be placed in wire gauze cages after attachment and hung in a live box.

In sponges in general, fertilization is internal and the egg develops in the body of the parent to the stage of the ciliated larva. Sponge eggs and embryos are commonly abundant in a breeding sponge and may be seen scattered through the interior with the naked eye. In some marine sponges asexual masses analogous to spongillid gemmules develop likewise in the body of the parent to ciliated larvae. In order to obtain larvae all that is necessary is to place a breeding sponge main aquarium Jar.

Growth of Sponges from Fusion Larvae

The ciliated larvae of *Lissodendoryx carolinensis* may easily be made to fuse with one another after they have begun to creep over the bottom of the culture dish and are thus approaching the phase in which they attach. It is only necessary to bring them in contact, coaxing them together with needle and pipette in a deep, round watch glass. The compound larva so produced has a feeble locomotory power. Using pairs that are nearly motionless, fusion masses of desired shapes may be produced on cover glasses. Or small excavations may be made in paraffin-coated dishes, and the larvae driven into such holes in large numbers. In this way, cake-like masses may be produced measuring 3-4 mm. in diameter. The smaller compound masses metamorphose without difficulty. The larger in the actual experiments died, sometimes after a partial metamorphosis.

3

Culture of Coelenterates

HYDRAS

Collection

Hydras most commonly occur attached to the submerged vegetation, fallen leaves, or other objects in pools, ponds, lakes, and the slow portions of rivers. They are collected from such habitats by gathering a quantity of vegetation and placing it in jars with a relatively small amount of water. As the vegetation begins to decay, the hydras usually come to the top of the jar or the surface film and may be picked out and transferred to suitable containers. One species, *Hydra littoralis,* occurs in enormous numbers on the under surface of stones in streams, spillways, and along shores subject to wave action. When the stones are turned over, this species appears like an orange jelly on the stone. They may be obtained in large numbers by squirting the animals from the stone into a pan by means of a squirter made from an atomizer bulb and a glass tube. Unfortunately, this species is not very suitable for laboratory cultivation, but is useful when large numbers are needed for a short time, or when large quantities of hydras are wanted for preservation. Hydras are sometimes found in large numbers attached to the surface film and in such situations are easily gathered. They do not in general live in stagnant water but require rather clean water with adequate oxygen supply.

Laboratory cultivation

The most suitable hydra for laboratory cultivation is the brown hydra, *Pelmatohydra oligactis,* but any of the species which live in standing water may be grown successfully in the laboratory. The species which inhabits moving water, *Hydra littoralis,* mentioned above, may be grown in the laboratory if the water contains a supply of algae or other vegetation to keep up a high oxygen content. The author has not tried bubbling air through a culture of this species but such a procedure would probably be successful. The green hydra, *Chlorohydra viridissima,* is one of the most hardy hydras, very common everywhere, and easy to maintain in the laboratory except for one difficulty. It is difficult to find a food crustacean which is small enough to be ingested by the green hydra. Cultures of the green hydra should be exposed to the light, but for other species a moderate light is best.

The first prerequisite of laboratory cultivation of hydras is suitable water. Hydras should never be put into freshly drawn tap water and in general only natural pond or river water should be used for their culture. If tap water must be used, it must stand for several weeks with growing algae or other aquatic plants in it so that it may become conditioned. The suitability of water for hydras should be tested by placing a few specimens in it. If these expand fully, with tentacles extended to their maximum extent, the water is suitable. In unsuitable water, the column remains contracted and the tentacles fail to expand. The presence of plants is not necessary in a hydra culture except in the case of *Hydra littoralis,* as stated above.

When a suitable water has been found, the hydras are placed in it, and fed daily. At intervals, the accumulation of bottom debris should be removed and small amounts of suitable water may be added from time to time. In general it is desirable that the cultures be covered.

In general hydra cultures succeed better if the temperatures are not too high; 20°C. is a very suitable temperature. The brown hydra, P. *oligactis,* is more susceptible to rise of temperature than any other species and usually dies at 25°C.

The great difficulty in the continuous culture of hydra is the occurrence of the phenomenon of depression. In spite of every care, hydra cultures will pass into this state at intervals and, unless prompt measures are taken, will die out. In depression, column and tentacles fail to expand, the animal ceases to feed, shortens to a stumpy appearance, and finally gradually disintegrates from the tips of the tentacles aborally. Depression is caused by over-feeding, fouling of the water, too high temperatures, and general aging of the culture with accumulation of waste products. The most successful method of reviving the animals from the depressed state is to transfer them to a fresh jar of suitable water. Lowering the temperature is also of assistance.

Sex organs when wanted for class display may be induced in most species of hydra, notably in the brown hydra, by placing the culture in a refrigerator for two or three weeks at a temperature between 10 and 15°C. Such cultures should be fed regularly. In nature most species are found with sex organs in the late fall but the green hydra is said to be sexual in the summer.

Food

Naturally for the continued maintenance of hydra in the laboratory it is necessary to have a food source that is easily cultured. The most suitable animal for this purpose is Daphaia because of its slow movements, weak resistance, and habit of moving about continuously. Some other cladocerans such as Simocephalus are easily cultivated also but are too strong and powerful for the hydras or else tend to stay on the bottom out of reach. Hydras will not eat ostracods. Oligochaete worms are eagerly accepted but their habits are such that the hydras would seldom have a chance to catch them. The very tiny newly hatched Daphnia in a Daphnia culture may be used as food *Chlorohydra viridissima;* some small forms such as Ceriodaphnia or the bosminids are suitable and may be grown like Daphnia. In case of necessity hydras may be fed on oligochaetes such as naids, tubificids, or enchtraeids, and often these may be purchased from pet shops. It is necessary to cut these up into pieces and to place the pieces in contact with the tentacles with a dropper or forceps; otherwise they fall to the bottom where they are out of reach of the hydras. This method of feeding is very time-consuming, but may be used when it is desired to save valuable experimental material. Experimental material which does not feed well of its own initiative, may often be fed successfully by placing a crushed daphnid, Cyclops, or bit of oligochaete in contact with the tentacles.

Tubificids are easily maintained in the laboratory for food by placing them in containers having two or three inches of pond mud on the bottom. For food, almost any kind of organic material such as boiled lettuce leaves, boiled wheat grains, pieces of bread, or bits of animal flesh, may be added from time to time.

REARING THE SCYPHISTOMA OF AURELIA

The scyphistoma of Aurelia proves to be a very hardy marine form; it may be maintained alive and in fairly active state of budding by keeping it in shallow dishes of seawater (it is well to have the seawater slightly hypotonic) and feeding with ground shrimp or particles of meat. Of course the water should be changed after feeding. The scyphistoma does best with a mixed diet. Scyphistomas have been reared at marine laboratories, using plankton or sea urchin ovaries.

4

Culture of Planaria

Planaria for use in nutrition experiments are collected in the field by placing small lumps of fresh liver in shallow water at the edges of ponds or streams where they are known to be present. The gather on the liver in a short time and may be rinsed off into collecting jars. They should not be crowded in the jars or they will be dead before the laboratory is reached. Likewise, they should not be crowded in the laboratory containers. For stock containers we use white enamelled milk pans, because in these the worms may easily be inspected to determine their condition. A city water supply containing chlorine is hot to be trusted. For some time it may appear to be harmless but when one observes that the worms are more restless than usual, that, all the worms in a container are in motion for an extended period of time, one should suspect that the tap water contains too much chlorine. The restless stage is followed by one in which the worms secrete large quantities of mucus and roll away from contact with the container. They gather in writhing masses and will disintegrate if they are not put into chlorine-free water. We use river or well water collected directly from the source.

After the worms have been distributed in the stock pans they must be treated as though in quarantine for about a month. Every day they must be inspected carefully and any worm showing the slightest irregularity in outline or surface texture must be removed. We have found that the worms come to the laboratory infected with parasitic diseases which develop quickly under the abnormal conditions of a laboratory environment. These diseases are capable of spreading and of aninhilating a large part of the stock. If one is to have stock reliable enough for experimentation, all disease must be eliminated, and with care this is easily accomplished. We always boil all water to be used on the worms in order to avoid the introduction of any disease-producing parasite. If at all possible, the water used should be perfectly clear because we have found that even slightly muddy water reacts unfavourably on the worms.

The laboratory routine in the care of the stock is as follows. The worms are washed every other day. This is done by plunging the hand into a lysol solution and then rinsing until no odor of lysol remains, for very slight amounts of lysol are highly poisonous for the worms. Then with the fingers the pan is wiped over its whole surface to loosen the dirty slime which always gathers. The worms settle at once to the bottom and all the water is poured away and replaced by fresh water. If the pan does not appear clean, this is repeated. Once a week the worms are washed into freshly sterilized pans and the dirty pans are thoroughly cleaned and sterilized.

Our stock worms are fed exclusively on raw liver and they continue in vigor and health for an indefinite period of time. The source of the liver must be considered because not all liver has correct nutritional value for the worms. For example, rat liver is poor food while beef liver is almost always excellent. We use liver taken from freshly-killed guinea pigs which are in prime condition. This has been found the best stock feed we have tried because we can control the diet of the animals furnishing the liver, and this is the determining factor in the production of nutritionally correct liver. If the stock is merely being maintained it is sufficient to feed once a week. The feeding is done by placing a small piece of liver in each stock pan. The worms feed readily and the liver is left with them for 3 or 4 hours. It is then removed and the stock pans are thoroughly washed. If one wishes to develop the stock rapidly, the worms should be fed twice a week.

To rear new worms for experimental purposes it is of course only necessary to cut the stock worms into pieces of suitable size and to allow time for regeneration. We always cut off the posterior extremities of the worms, the length of the piece cut off depending upon the size of the worm. We separated these tail pieces into pans by themselves and allow them to stand without feeding for a period of 4 weeks, at which time they have attained the adult shape and are ready for nutritional experiments.

We keep our experimental worms in an incubator with a

temperature of 24° C., but the stock and regenerating worms are kept in the laboratory with the heat turned on at night during the cold season. If the worms become thoroughly chilled many of all or them will die. This happens above the freezing point and makes it wisest to keep the stock pans away from cold windows.

PLANARIA

For eight years a culture of *Planaria [=Euplanaria] maculata* has been kept under observation and fed very successfully with enchytraeid worms. The animals were collected in a large pond and a selection was made of the individuals which would accept the enchytraeid worms as food.

The planaria are kept in a wooden tub (24 inches in diameter and 11 inches deep), charred on the inside, and are fed about once a week. Approximately 2000 individuals are maintained in this space. When fed, a level teaspoonful of worms is dropped over the bottom of the tub. Several planaria will cluster over each worm and so share the food. The amount of food given is regulated by the growth of the planaria: if they are getting smaller, more enchytraeid worms are offered; if more individuals are needed, additional food will speed up their growth and reproduction.

When large numbers of planaria are needed for class material, we are careful not to disturb the tub for two or three days. A film then forms over the surface of the water. To bring the planaria to the surface, the sides of the tub are tapped with a hammer. Each animal is then picked out with a fine glass rod which is slipped under its dorsal side as it floats ventral side up. The planaria folds its dorsal surface around the rod and it is then dropped quickly into a dish for study. Without the film on the surface of the water the planaria will not stick to the rod.

Different species of planarians live in different sorts of habitats. Some, notably our most common species, *Euplanaria maculata,* live in ponds, lakes, and the slow parts of rivers on the vegetation and on the under surface of stones, leaves, or other objects. They may be obtained by turning over stones and fallen leaves and washing the animals into a pan by means of a strong squirter made of an atomizer bulb and glass tube. Their presence on vegetation may be ascertained by shaking small samples of the vegetation in a vessel of water. If they are present large quantities of the vegetation should be gathered and placed in pans with a small amount of water. As planarians cannot endure stagnant water, they soon come to the top and may be picked off.

Other species, notably the large dark forms such as *Euplanaria agilis* and E. *dorotocephala,* live in spring-fed streams and marshes. Their presence may be discovered by baiting a suspected habitat with a piece of raw meat placed along the edge, not in the current. After 15 or 20 minutes, the piece of meat should be turned over and planarians, if present, will be found attached to the under side. The entire habitat should then be baited with meat. Fresh raw beef is best and should be cut into pieces 1 or 2 inches wide and 2 or 3 inches long. Such pieces should be distributed throughout the edges of the habitat so as to rest partly in the water, partly above the water. At intervals of 15-20 minutes the pieces should be picked up with a long forceps and shaken off into a jar of water. With a few trials of this sort the best spots in the habitat are soon discovered and all of the meat may be moved to such spots. In preparing such collections for transport back to the laboratory, they should be washed free of bits of meat by several rinsings, and the jars filled not more than ¾ full with fresh clean water from their habitat. A depth of not more than an inch of planarians should be allowed to a pint jar. In bringing them in, care must be taken to avoid high temperatures. The jars must not be set on the floor of a car which is apt to become hot from the engine.

Baiting with meat is usually ineffective with pond habitats and commonly succeeds only with species which live in flowing water. Some species, however, even in flowing water, respond poorly to this method and must be picket from stones and water weeds by hand. Among the species which the author has personally seen or knows may be collected successfully by baiting are: *Euplanaria agilis,* E. *dorotocephala, Fonticola velata,* and *Phagocata gracilis. Euplanaria maculaia* and *Procotyla fluviatilis* usually respond poorly to meat baits.

Laboratory maintenance

Planarians of practically any species may be kept successfully in the laboratory in glass or crockery containers or enamelled pans. These should be darkened by means of suitable covers. Treated city waters are not very suitable, but most species will live in such water for a considerable time. Spring or well water is desirable.

Those species mentioned above as collectable by baiting with meat are also the ones which may be kept most easily and successfully in the laboratory .They are fed two to three times a week on beef liver (pig liver is not suitable). Before feeding, the water in the pan should be lowered to a depth of several inches. The liver should be cut into long thin strips and disposed over the bottom. The pan is then covered and left undisturbed for 2 or 3 hours, after which the liver is removed, the pans thoroughly rinsed, and filled with fresh water. Even if the animals are not fed, the water should be changed two or three times weekly as planarians are very susceptible to fouling of the water. All food fragments should be carefully removed. E. *dorotocephala,* E. *agilis,* and *Curtisia foremani* are very easily kept by this method for long periods of time in the laboratory; *Fonticola velata* and *Phagocata gracilis* may also be maintained on liver, although not so well as the first-named species. In place of liver, pieces of earthworm, clam, etc., may be used; some forms prefer such food. Yolk of egg drawn out with a dropper into a strand on the bottom has been employed successfully.

E. *maculata* does not feed very well on beef liver and is less easy to maintain in laboratory culture than the preceding species. It is necessary with this species to grind the liver in a meat grinder and wash it thoroughly in running water. Small bits of such washed liver will usually be accepted as food. This species, however, in general prefers pieces of invertebrate flesh or liver or other flesh of tadpoles, fish, *etc.*

Procotyla fluviatilis is the most difficult of our common species to keep under laboratory conditions as it will eat nothing but live prey, such as daphnids, amphipods, and isopods. It will sometimes accept blood clots, but in general it is impractical for laboratory purposes.

Sexual material

Zoologists at times desire sexually mature material for class or experimental purposes. In general those species which reproduce extensively by asexual methods are seldom found in the sexual state; this statement applies to E. *dorotocephala,* E. *agilis,* and *Fonticola velata.* E. *maculata* and its various varieties are commonly sexual throughout the summer time and numerous egg capsules will be found on the under side of the stones in the habitat of this species. It is a curious fact, however, that E. *maculata* apparently never sexual in some localities or regions while always sexual in the summer in others.

Sexual specimens will continue to lay eggs under laboratory conditions.

Species which do not reproduce asexually are commonly in the sexual state at some definite season of the year. For *Procotyla fluviatilis,* which, owing to its transparency, is our most suitable species for preparing slides showing the reproductive system, the time of sexual maturity extends from September into the winter or even, in some localities, into spring. *Fonticola morgani (=Planaria truncata)* is sexual in summer, as is also *Polycelis coronata* of mountain streams of the northwestern U.S.

The only species which may be depended upon to lay egg capsules regularly under laboratory conditions is *Curtisia foremanii (=Planaria simplissima)*. This species occurs throughout the Atlantic coast states, may easily be cultivated in the laboratory on beef liver, and will lay egg capsules continuously for a long period. The young soon grow up to sexual maturity and also lay in their turn so that a continuous supply of capsules is assured with this species.

5

Culture of Annelids

NEREIS

Nereis limbata, which occurs abundantly in Eel Pond is the one species in this vicinity which is known to have a distinct and unmistakable lunar periodicity in spawning. Eggs may usually be had in great abundance roughly from the full moon until new moon during all of the summer months and not to any considerable extent at any other time.

The eggs and spermatozoa are extruded at night from 9 to 10 p.m. as the sexually mature worms swim at the surface of the sea. As the males and females come into contact with each other they are stimulated to expel their gametes vigorously. The stimulus, however, is chiefly chemical rather than physical. At this time in the month the body cavities of the worms are distended with eggs or spermatozoa and after they are expelled nothing but the ghost of a worm remains.

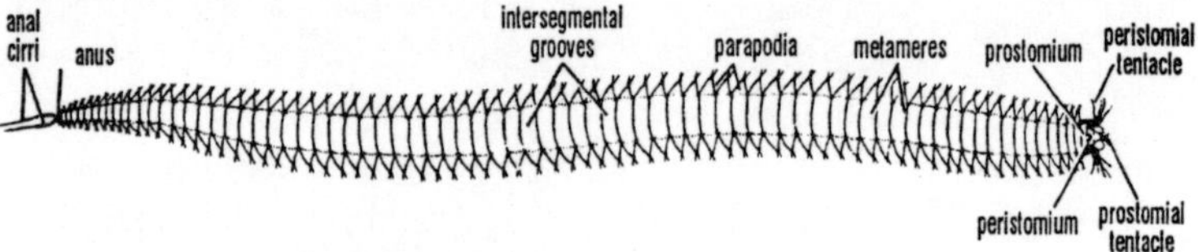

Fig. Nereis. External features

Method of Collection

A small dip net and a lantern or flashlight are needed in collecting. The worms are attracted to light and may be dipped up and placed in suitable dishes of seawater. Females should be kept separate from males, otherwise they spawn at once.

Method of Securing Eggs

Select a distended female, place her in a clean dish of seawater, and with scissors cut across her body to allow the eggs to escape. In the same way cut a male in two in a dish containing 25 or 50 cc. of seawater. After washing the eggs once or twice by pouring off the water and refilling the dish with fresh seawater, add three or four drops of spermatozoa and agitate gently. Within five minutes after insemination the eggs extrude a jelly in which they lie embedded. The eggs are thus readily fertilized artificially and development proceeds. Usually 100% of the eggs cleave and, barring accidents, nearly all develop into normal embryos.

Care of Cleaving Eggs

The cleaving eggs require no further attention except that the water should be changed several times during the next 12 or 15 hours. In the meantime the eggs cleave and acquire cilia. After 24 hours the embryos may be separated from the jelly and transferred to a clean dish of seawater either by pouring or by using a wide-mouthed pipette. Care should be exercised to get rid of all decaying organic matter as soon as possible, and this must be accomplished within 36 hours after insemination or before. Cleaving eggs that are allowed to develop without frequent change of water usually develop abnormally or die.

Schedule of Development

The embryo becomes an early gastrula 12 or 15 hours after fertilization, with the four large macromeres constituting the principal part of the endoderm. It is ciliated and rotates in the jelly. It is a late gastrula after 24 to 30 hours, the rate of development depending upon the temperature. Between 36 and 48 hours the larva is a trochophore, at first spherical but later somewhat elongated. On the third day the first three segments of the worm body are completed and no additional segments are added for several days although the embryo increases in size. It is possible to keep the embryos until other segments grow but to do so requires special feeding methods. The Nereis larva is unusually hardy and easily cared for. The trochophores and early segmented larvae are active swimmers but as the ciliary mechanism becomes inadequate they depend more and more upon wiggling and creeping.

Care of Larvae

The egg of Nereis is large and well supplied with yolk and oil so that the larvae require no feeding during the first five or even seven days of development. They are easily cared for because after two days they have a tendency to settle to the bottom on one side of the dish and may be transferred to a clean dish of seawater with a wide-mouthed pipette. This should be done once per day or more frequently in hot weather. After five days they may be fed upon diatoms but if it is desired to keep them for several weeks it is best to transfer them to a large cylindrical balanced aquarium, containing a dense culture of developing diatoms which adhere to its sides. Just reared Nereis megalops to maturity in such a jar of diatoms.

A Method for Rearing Nereis

The writer has reared two species of Nereis to sexual maturity in the laboratory by the following method. One species, *Nereis agassizi*, was reared to sexual maturity during each of two consecutive years and in the third attempt the larvae were maintained for a period of nearly 14 months but due to unfortunate circumstances did not reach sexual maturity. A second species, *Nereis procera*, was successively cultured in the laboratory for a year. At the end of that period the worms had reached sexual maturity.

While the worms are "swarming" at the surface in their seasonal spawning it is a comparatively easy matter to capture both males and females. Better results may be obtained if the males and females are kept separate during capture and transfer to the laboratory. They should spawn in separate dishes in sufficient seawater to keep them well covered and to maintain a fairly even temperature. A small amount of water containing spermatozoa is added to the eggs and thoroughly mixed. Care should be taken not to use too great excess of spermatozoa. The dish containing the fertilized eggs should be allowed to stand for a few minutes and then be emptied into a larger container (battery-jar) containing 2 or 3 litres of seawater. This jar is placed in running water or in a suitable location to maintain a fairly constant temperature. The degree of temperature best suited to a particular species would seem to be that of the environment from which the adult worms were taken. After the eggs have settled to the bottom, as much of the water as possible should be siphoned off to remove the excessive spermatozoa from the culture and fresh seawater should be added. Polar bodies begin to appear after 1 to 1½ hours and cleavage starts after 2½ hours. The cleavage rate is usually fairly rapid and movement of the larvae begins in 12 to 15 hours. The trochophore stage is reached in about 36 to 48 hours and larvae with three pairs of setigerous appendages appear in between 3 and 4 days.

It is highly important that the temperature be kept constant. The water should be changed daily and agitated at least once each day to provide aeration. When the larvae have developed setigerous appendages they will soon be provided with jaws and are then ready to begin feeding. The larvae will consume very readily small diatoms which are then added to the culture. For this purpose it was found that species of Navicula and Nitzschia were of suitable size. They were eaten in large quantities. The larvae grow rapidly and when they have developed 6 to 8 segments they form mucous tubes within which they hide themselves. When the larvae reach an age and size having 10 or 12 segments they will consume other food in addition to diatoms. At this time Ulva and brown kelp, such as Nereocystis, may be added to the diet. At this time they may consume other small worms and any material that they can devour .The worms extend themselves almost their entire length from their tubes and reach for food.

After the young worms develop tubes it is not necessary to change the water so frequently. A change every 2 or 3 days is sufficient, and at a later date, once a week will suffice, provided there is plenty of green vegetation in the water.

Several important factors must be kept in mind :

(1) there must not be overcrowding of organisms;

(2) fresh water must be provided to allow for a sufficient supply of oxygen;

(3) care must be taken to prevent contamination;

(4) the temperature must be kept within a limited range; and

(5) a sufficient quantity of suitable food must be provided at all times.

Culture of Hydroides lexagonus.

Hydroides hexagonus, a serpulid worm, secretes a calcareous tube which adheres firmly to shells of mollusks, stones, and wooden structures. The breeding season opens between June 10 and June 15 and close between October 15 and November 1.

Method of Obtaining Eggs and Spermatozoa

To obtain eggs or sperm in it advisable to remove the worms from their calcareous tubes and place them in stender dishes of Syracuse watch crystals filled with seawater, one worm per dish. When so treated they always spawn immediately if they are sexually mature. The sexes are separate and the gametes, whether eggs or spermatozoa, are carried free in the coelomic cavities from which they are extruded through nephridiopores located along the sides of the body. During the early part of the breeding season over 50% of the spawned eggs are immature and undersize. Later on they are nearly all mature and fertilizable. Maturation does not take place until the spermatozoon enters the egg, the germinal vesicle being indistinctly visible through the yolk.

Embryology

It is desirable to delay insemination of the eggs for half an hour after spawning has occurred because the spermatozoa are quite immobile when first extruded. Under the stimulus of seawater they gradually become activated but are relatively inactive at best. After allowing time for activation, remove the eggs by means of a pipette to a fresh dish of seawater and add four or five drops of sperm.

After fertilization the eggs develop within ten hours into actively swimming gastrulae and therefore rise from the bottom. They may now be poured into a clean dish, thus discarding the eggs which failed to develop.

Within 24 hours the embryos have become transparent trochophore larvae which continue to swim actively. They remain in the trochophore stage of development for 10 days or two weeks showing little external change except a slight slender outgrowth at the posterior end which constitutes the beginning of the worm body. The feed readily upon diatoms by means of ciliary mechanism and may be kept indefinitely under laboratory conditions.

Because they at no time settle to the bottom, it is difficult to keep the water changed but they remain in good condition if poured daily into clean dishes discarding the bottom layers. They have a tendency to collect at one side of the dish and may be transferred to clean dishes of seawater by means of a pipette, but this method involves the loss of many embryos. Zeleny has reared them to metamorphosis in aquarium jars.

Post Embryonic Development

The trochophore finally develops a slender worm body, settles permanently, and secretes a calcareous tube. By placing mollusk shells or stones in a cage and sinking them in shallow water which is known to contain breeding Hydroides worms it is possible to secure many young worms and study their further development. Studies of this character carried on at Woods Hole for several years have shown that they become sexually mature in seven or eight weeks before they are half grown. They become fully grown in two years. Most of the worms ordinarily collected are only one year old and it is likely that many if not most of them die during the second year.

The shell of an average-sized worm after one year's growth measures 65 or 70 mm. in length and 3 or 4 mm. in widest diameter. The largest worms may reach 120 mm. in length and 5 mm. in greatest diameter.

CULTURE OF SABELLARIA VULGARIS

Sabellaria vulgaris is a sedentary polychaete found along the Atlantic coast from North Carolina to Cape Cod.

The worms live within sand tubes which they build on stones, empty sanddollar and oyster shells, and occasionally on Bryozoa modules and Limulus shells. Males and females are about equal in number in the collections. The sexes can be recognized externally only in those fully mature animals which contain a large number of either eggs or sperm. The abdominal segments, which are greatly distended, are dense white in the male and a decided pink in the female. The eggs or sperm are shed almost immediately after the animals are removed from their tubes. The number of gametes shed is greated from those animals which have a pronounced colour in the abdominal segments, but even animals which show neither the distinct white nor pink colour may shed abundantly.

Animals collected throughout the greater part of the summer showed no apparent differences in the condition of their gametes. Eggs from worms dredged at irregular intervals during the periods. August 22 to September 7,1934 and June 24 to September 9,1935, developed normally in more than ninety-five per cent of the cases. The same high percentage of normal development usually followed from eggs of animals that had been kept in aquaria with running seawater for as long as nine weeks.

Obtaining Eggs and Sperm

Uninjured animals are most easily obtained using the following procedure; (1) remove the sand tube from the shell or rock, (2) break away enough of the tube to make the head and tail visible within, (3) carefully force the animal from the tube by inserting a blunt probe into the head end of the tube.

The sex of the animal removed from the tube is ascertained, if the colour of the abdominal segments is not definitely white or pink, by placing it in a few drops of seawater until if begins to shed. The sperm usually pour out of the male in dense white clouds. The masses of eggs shed by the female break up in the water into small groups. The male is placed into four drops of seawater until it has completed shedding. The female is placed into a finger bowl containing about 200 cc. of clear seawater. After it has shed for a few seconds, it is moved to a new position in the finger bowl and allowed to shed in that place for the desired length of time, depending on the number and kind of eggs desired. The first eggs to be shed are not generally used because of the possibility of their having been on the surface of the animal while exposed to the air.

When preparing the eggs for fertilization, it is best to allow them to remain in the seawater for about fifteen minutes. Towards the end of that time, the sperm suspension is prepared. One drop of the sperm shed into the four drops of seawater is diluted with four drops of seawater, and one drop of this diluted suspension is then added to a finger bowl of seawater (about 260 cc.). The eggs are drawn up with a narrow medicine dropper and transferred into this suspension.

The original sperm suspension (made by allowing the male to shed in four drops of seawater) may still be used after three to four hours and longer if evaporation of the water is prevented. The eggs may be fertilized immediately after shedding, but it is best to allow them to remain in the water for about fifteen minutes before they are fertilized, if eggs of the same stage of development are desired.

Early Development

Just after being shed, the egg is very irregular in shape and contains a large clear germinal vesicle. A few minutes later, the egg begins to round out, a clearly visible membrane is raised from the surface, and the large germinal vesicle breaks down to form a spindle which extends across not quite half the diameter of the egg. The average diameter of the rounded egg is fifty-six micra; the membrane, usually wrinkled, is about twelve micra from the egg surface. Fine protoplasmic processes are seen to extend from the egg surface as the membrane is raised. The elevation of the membrane is apparently due to the swelling of a rather dense jelly situated between the egg surface and the membrane. This jelly is ordinarily invisible but it can be demonstrated by removing the outer membrane and placing the eggs in a dense suspension of Chinese ink.

The process of fertilization has been described by Waterman and will be further discussed by me in a future publication. The sperm attaches to the egg within the first minute after mixing the eggs and sperm, but the exceedingly large fertilization cone may not be completely retracted for a period of twelve to fourteen minutes. Some time during the course of sperm entry, the protoplasmic processes are withdrawn into the egg. The egg is too opaque to allow for the direct observation of any internal processes other than those connected with the spindle area.

Preceding the formation of the first and second polar bodies there is a distinct flattening of the egg at the region where the polar bodies will be extruded. The first polar body separates at about nineteen to twenty-three minutes after fertilization; the second comes off nine to eleven minutes later. At fifty to fifty-five minutes after fertilization, a large anti-polar lobe is formed and the cell divides into two (sixty-five to seventy minutes). After the division the polar lobes goes into the CD cell. Ten to fifteen minutes later, a smaller lobe is given off at the anti-polar end of the CD cells and the second cleavage occurs. At the completion of the division the lobe flows into the D cell. The first set of micromeres, which are only slightly smaller than the basal cells, comes off in the usual dexiotropic fashion. During the course of this division a third lobe forms in the D cell. The later cleavage has not yet been described.

At five and a half hours, the developing embryo begins to move about by means of cilia and at eight hours the apical tuft and prototroch are well formed. The larvae live fairly well on a diet of the diatom Nitzschia. One larva was raised, without much care, to the beginning of metamorphosis (seven weeks) at which time it was fixed. Wilson has given a detailed description of the larvae of the British Sabellarians with which those of *Sabellaria vulgaris* agree very closely.

The exact time relations in development and the effect of change of temperature on such relations have not been studied. The schedule of events as given above is only approximate for room temperatures varying from nineteen to twenty-five degrees C.

Removal of Outer Membrane

The outer membrane of the egg may be removed by following a slight modification of a procedure described by Hatt. Two solutions A and B, are prepared as follows :

Solution A.-A given volume of a solution of 1 gm. of Na_2CO_3 in 1000 gm. of isotonic NaCl (or KCl). This gives a solution with pH 9.6.

Solution B.-0.45 cc. of 1.0 N HCl are added to 100 cc. of isotonic NaCl (or KCl). This solution has enough HCl so that when an equal amount of the solution is added to a given amount of Solution A the resulting mixture is at the pH of seawater (8.2).

The eggs are placed in a Syracuse dish and as much of the seawater as possible withdrawn, without injury to the eggs. A small amount (three to four cc.) of Solution A is poured on them and the dish rotated. After about one minute, the eggs are allowed to settle and the solution withdrawn. Now a carefully measured amount (five cc.) of Solution A is pipetted into the dish. The eggs clump into large masses, but as the membranes dissolve the eggs separate and the masses break up. The disappearance of the egg membranes may be followed under the microscope. When they have disappeared from most of the eggs (about five minutes) five cc. of Solution B is quickly added. The force of the stream from the pipette is sufficient to mix the solutions. After the eggs have again settled, they are transferred to fresh seawater in Syracuse dishes. The bottoms of the dishes have been previously coated with a thin layer of agar to prevent the adhesion of the eggs to the glass surface.

This treatment removes the outer membrane but the jelly still remains about the egg.

CULTURE OF TUBIFICIDAE

Oligochaete worms of the family Tubificidae form an excellent food for many kinds of laboratory animals including planaria, leeches, dragonfly and damselfly nymphs, aquatic beetle larvae, and many fishes. A star-nosed mole kept in the laboratory ate them voraciously. The ease with which these worms may be collected in quantity and kept for months in the laboratory, and the avidity with which they are eaten, make them important laboratory animals.

Tubificidae occur in a considerable variety of freshwater habitats. They may be found most readily and most abundantly in the mud, or in muddy borders, of streams or ponds where considerable organic matter is undergoing decay. When in shallow water they may often be seen with their tails waving in the water, their heads buried in the mud. The location of tubificids in soft muddy borders of streams or ponds may often be determined by noting numerous small casts on the surface. When their presence is suspected take up a small quantity of mud on a trowel and examine it for worms. If they are abundant determine how deeply they are embedded in the mud, then scrape or scoop up the layer of mud containing them with trowels or small shovels and put in 10- or 12-quart pails.

At the laboratory put the contents of a 12- quart pail in a shallow galvanized pan measuring about 15 x 12 x 3 inches deep. Set the pan on a drain table in a cool room and allow a very small stream of water from a faucet to flow continuously through a rubber tube into the pan. The worms will come to the surface in a few hours. During the night small masses of worms tend to migrate out of the pan and onto the drain table if the pan is overcrowded. These may be used for feeding purposes until the stock is reduced.

If the mud rises in the pan because of the formation of gas, prick holes in it and press it down.

In quiet rivers receiving sewage or in ponds to which manure has been added to increase productivity of fish food, tubificids often collect in masses as large as a man's fist, or larger, at the surface, either on or near the decaying material and frequently near the margin or even upon the muddy border. They occur when the water is warm and disappear when it gets cold in the fall. Such worm masses are collected with a large tea strainer, dipper, or long-handled fine-meshed dip net. In the laboratory they may be put with mud and decaying vegetable matter, or with manure, potato, or other food material free from mud.

To feed tubificids the following materials have been used; fresh horse manure, baked potatoes cut in halves, boiled potatoes, butts from head lettuce, masses of bran, and bread. Press food down into the mud. The horse manure and bran should be buried in the mud.

The worms are removed in small masses by means of a pair of forceps. Wash them to remove mud before feeding them to other animals.

CULTURE OF EARTHWORMS

Since living earthworms are useful in the laboratory for demonstrating behaviour, and since freshly killed earthworms are far superior to preserved specimens for the study of certain organ systems, especially the digestive, circulatory and excretory systems, many laboratories need a supply of living worms for mid-winter use. Brief consideration will be given to the culturing of three species.

Specimens of *Lumbricus terrestris,* the common earthworm of the United States, must be gathered at night and preferably between 10:00 and 12:00 O'clock during or following a drizzling, warm rain when the ground is thoroughly soaked. The worms come to the surface of the ground in large numbers at such times and may be captured easily with the assistance of a strong flash light or an acetylene lantern. The best collecting ground are closely cut lawns where the soil is rich.

When the worms have been collected they may be left for the remainder of the night in a cool place in a pail containing a small quantity of freshly cut grass. The next morning the worms should be carefully sorted, and all injured or abnormal specimens should be removed. If they are washed and placed, a few at a time, in a dish of water those that are injured may easily be detected.

Earthworms feed very largely on dead and decaying leaves and, like chickens, they digest their food better if there is a certain amount of grit in their diets. We have obtained best results by keeping earthworms in large boxes filled about 12 inches deep with approximately equal parts of old leaves and leaf loam gathered in the woods. Under no conditions should heavy clay soil be used. The worms need no other food, as they feed on the dead leaves. The material should be kept moist but not saturated with water. Unless extreme care was exercised in removing all injured worms, the boxes should be inspected after a week and all dead and dying worms removed. Should it happen that the worms are not keeping well those that are healthy should be removed and placed in a fresh box of leaves and loam.

Earthworms also keep well in very light, loamy soil. If this is used it is often advisable to feed the worms. Bread crumbs or corn meal make excellent food. The food should be moistened with water, spread sparingly over the top of the soil every 2 or 3 weeks and covered with about an inch of loam. Feed sparingly and not too often or the food will spoil and the worms may die.

Avoid trying to keep too many worms in one box. A cubic foot of culture material, after it has settled, will be sufficient for about 50 worms. Cover the boxes with panes of glass and keep cool. Temperatures above 60° F. usually prove fatal.

While cocoons of this earthworm are not easily obtainable, a few of them may usually be found by carefully sorting over the loamy material in the boxes after the worms have been stored in it for a month or so. The young worms emerge from the cocoons in a few weeks and thrive under the same treatment as that given the adults.

The fecal earthworm, *Eisenia* [= *Allolobophora*] *foetida*, which is much hardier than *Lumbricus terrestris*, and which is rather extensively used for experimental purposes, keeps very well in partly rotten cow and horse manure. The worms may usually be found here in abundance. They copulate and form large numbers of cocoons in the laboratory, if kept in containers supplied with this material.

The small white earthworm, *Enchytraeus albidus*, lives well in the laboratory. In addition to being an excellent food for small fish and Amphibia it may be narcotized with chloretone and used to demonstrate many annelid structures with the aid of a binocular microscope. Keep in boxes filled with black loam and feed sparingly with bread soaked in milk or water. After each feeding the food should be covered with about an inch of loam. The material should be kept moist but not soaked. The worms prefer temperatures around 55°-60° F. Higher temperatures should be avoided.

6

A Culture of Daphnia

Fleischmann's yeast has been fed to a mass culture of *Daphnia magna* with striking results, reproduction and growth being markedly more rapid, and population more dense than with any of the usual media.

About ¼ of a fresh yeast cake is mixed into a uniform suspension with from 50 to 100 cc. of water, and poured into the aquarium, which contains from 60 to 70 litres of water. The feeding is repeated every 5 or 6 days. It is necessary to have a stream of air bubbling through the medium at all times, or the yeast may prove lethal, probably by giving off CO_2.

The method has not been tried on other species of Cladocera, except *Moina affinis*, with which it was equally successful, nor has it been tried with few animals in small containers, but it is so successful in the mass culture that it seems wise to make the food material known. It should be particularly useful in physiological work, in which the usual manure infusion may be a source of large quantities or unknown solutes. It should also be valuable in raising Daphnia in large numbers as food for other organisms.

Methods for Culturing Daphnia

1. *Algae Method. (a)* Run tap water into battery jars or butter tubs and allow to stand for 24 hrs. for the purpose of getting rid of air bubbles; otherwise the water-fleas adhere to these and are carried to the surface where they soon perish.

   (b) Add sufficient algae to tinge the water slightly green and inoculate with Daphnia.

(c) The alga *Coccomyxa simplex* may be cultured in large quantities by a method developed by H. Schomer. When the culture is at its peak it is centrifuged with a Birge-Juday centrifuge and it is this centrifugate that is added to the Daphnia culture. The algae may be grown in battery jars. The Schomer medium consists of:

KH_2PO_4 2.7 gm. per litre

$MgSO_4$ 4.9 gm. per litre

$Ca(NO_3)_2$ 4.7 gm. per litre

This method surpasses any I have tried in bringing about maximum rate of reproduction.

2. *Yeast Method* (Modified from Bond's). a. Same as

(a) above.

(b) Make a thick suspension of moist Star or Fleischmann's yeast in a flask by shaking chunks vigorously until a suspension forms; then add the suspension to the culture until the water is slightly milky. When the water-fleas have cleared it, add the same amount again. Water should be completely changed once a month and the crop seined often to prevent crowding. Aeration is not necessary, but the maximum reproductive rate is reached with aeration. Stirring once a day is sufficient to keep the culture going satisfactorily.

(c) Augmenting the yeast diet with the algae as described above is recommended. If the cultures are in good light, algae generally grow on the sides. They may be scraped off the walls with a razor blade and then broken up by stirring. These serve well to augment the yeast diet. One may also add about a teaspoonful of dry sheep manure per 4 gel, every two weeks. The addition of these substances gives D. *magna*, for example, the natural red colour, whereas if they are raised on yeast alone their colour is almost white.

3. *Banta's Method*. This method gives very satisfactory results. I raised Daphnia for 9 months in total darkness, using culture water made according to his directions.

4. *Sheep Manure Method. (a)* Let water stand in suitable containers 24 hours.

   (b) To every gallon of water, add 1 teaspoonful of dry sheep manure, 0.5 gm. of acid phosphate and a litre of aqueous soil filtrate (water that has been allowed to filter through rich garden soil).

   (c) Allow mixture to stand for a day and inoculate with Daphnia.

5. *Aquarium Water Method.* Aquaria that have become very green with a phytoplankton growth furnish a very convenient culture medium for Daphnia. Just remove the fish or transfer the water to another container and inoculate with Daphnia. This method was recommended to a tropical fish fancier who was having trouble with "green aquaria." He removed the fish, then inoculated with Daphnia; after the Daphina had multiplied sufficiently to "clear" the aquaria, the fish were put back and had a real feast!

20°C. is a satisfactory temperature at which to keep these cultures. Artemia may also be raised by any of these culture methods. I have kept individual Artemias living for 4 months by using algae. The salt content must be regulated (2 teaspoonfuls to 1/2 pint of water).

If it is desired to cease the Daphnia cultures during vacation periods, it is only necessary to chill or "crowd" the cultures and thus produce ephippial eggs. These may be collected and stored. To hatch them, place outside a window for two weeks of October weather so that they freeze and thaw several times, then place in water to hatch. They may also be artificially frozen in a refrigerator; 8 thaws and freezes give the maximum yield.

Propagating Daphnia and Other Forage Organisms Intensively in Small Ponds

For the first few years we tried to keep cultures going continuously throughout the summer in the same pond, merely adding from time to time a definite amount of fertilizer. The result was that very successful cultures were maintained during May and part of June which always ran out in July but seemed to come back in late September and October.

The most important causes for the diminishing supply seemed to be the population density, the accumulation of waste products not only from the Daphnia themselves but from micro-organisms also present, predatory enemies of Daphina, and probably water temperatures somewhat above 82°F. All of these factors have been mentioned before by various investigators, especially by Dr. A.M. Banta and his associates. All of the above mentioned factors except predatory enemies favour the production of so-called winter eggs, which is usually an indication that the culture is on the wane.

In order to have strong cultures at all seasons it becomes necessary to keep the individual Daphnia in the active condition of producing asexual or so-called summer eggs only, which demands the elimination of the inhibitive factors just mentioned.

A too dense population, of course, is easily reduced by using the Daphnia. The excessive accumulation of waste matter is corrected by a change of water in the pond. Predatory enemies are controlled partly by sterilizing the pond bottom and sides immediately before starting a culture and later by spraying the surface with some non-toxic animal oil, such as herring oil, salmon oil, or cod-liver oil. The water temperature cannot be controlled and, consequently, where it ranges above 82° F. for any length of time it may be difficult or impossible to produce cultures of *Daphnia magna.*

A population density reaches the maximum under the experimental conditions maintained in our ponds is from 16 to 26 days with the water temperature varying between 70° and 80° F. Consequently cultures are permitted to develop for 21 days, when they are fed to the fish and an entirely new culture with fresh water is started. The schedule of operation is as follows:

First day-Pond is drained, bottom and sides thoroughly disinfected with a strong solution of chlorinated lime, allowed to stand 6 hours; then refilled, fertilized, and stocked with large Daphnia from an active culture.

Fifth to seventh day-Second fertilization.

Tenth to fourteenth day-Third fertilization.

Twenty-first day-Drawing the pond and using the Daphnia.

In general this routine was continued through several summers from May to October and when a proper amount and kind of fertilizer was administered, the results were consistently good.

Although we have tried to maintain pure cultures of Daphnia, other organisms have naturally appeared and multiplied. Some of these are desirable food animals but others are predators.

Several little hard-shelled ostracods have appeared in considerable numbers, especially late in the culture period. We believe there is some connection between their abundance and a decline in the production of Daphnia, without as yet having direct evidence that they actually prey upon living Daphnia. Disinfection of the pond does not entirely eliminate them but nevertheless helps to keep them under control.

Midges of the genus Chironomus are attracted to the pond within a day or so after adding fertilizer, probably by the odors of fermentation, and deposit enormous numbers of eggs. These add considerably to the production of the pond. Mosquitoes likewise are attracted at about the same time and their characteristic floating egg masses become conspicuous all over the surface.

Table 6.1. Amount of Fertilizer per 100 Cubic Feet of Water and Its Apportionment

Fertilizer	*Soy Bean Meal*	*Cotton Seed Meal*	*Dry Buttermilk*	*Sheep Manure and Soy Bean Meal*
Initial Dose	1 pt.	1 qt.	½ pt.	4 qts. Sh. M. 1 qt. B.M.
2nd Dose	5th day – 1pt.	7th day – 1.5 pt.	7th day – ½ pt.	14th day ½ pt. B.M.
3rd Dose	10th day – 1 pt.	14th day – 1.5 pt.	14th day – ½ pt.	
4th Dose	15th day ½ pt.			

These three associated organisms may be considered beneficial insofar as they increase considerably the production of fish food. The mosquitoes, however, are obnoxious as adults and since they transform to the adult stage long before the Daphnia culture reaches a peak, it is probably better to exterminate them with non-toxic oil spray 4 to 6 days after the eggs appear . Other groups of animals which almost invariably appear are the back-swimmers, larvae of aquatic beetles, nymphs of dragonflies and damselflies, and hydra. All are predatory on Daphnia and midges. It is well known that the oil spray kill all insects that must come to the surface for air, such as the larvae of mosquitoes, beetles and adult back-swimmers. The dragonfly and damselfly nymphs and hydras are eliminated when the pond is drained and disinfected. Many different fertilizers have been tried, including manure from horses and cattle, dried sheep manure, acid phosphate, soy bean meal, cotton seed meal, dry buttermilk, and alfalfa meal. The alfalfa meal produced only fair cultures and required such a large quantity of material that experiments were early discontinued. The dried sheep manure used in combination with either acid phosphate or soy bean meal produced average cultures consistently. The wet animal manures were from ordinary barnyard piles containing much straw and slightly rotted. The resulting cultures were about average but not always dependable. The fertilizers which gave cultures averaging the highest were dry buttermilk, soy bean meal, and cotton seed meal. Very little difference between them was noted.

Peak cultures seem to require from 30 to 50% more cotton seed meal than soy bean meal and the culture period is somewhat longer than with the former. On the other hand, cotton seed meal, like animal manures, stains the water a deep brown, resulting in deeply coloured red Daphnia. The deep colour seems also to prevent undue growth of blanket algae. The soy bean meal produces light gray Daphnia and an abundance of free-moving micro-algae which colour the water green. It also encourages the growth of blanket algae.

The quantity of Daphnia necessary to start a successful culture, within certain limits, is not so important as the physiological condition of the mother organisms. They should be active summer egg producers. Almost always a few will be found with winter eggs. If the proportion is large, specimens from such a culture should be discarded. In the experiments reported here, from 25 to 100 cc. of mother organisms were generally used. We believe that 50 cc. is sufficient for 100 cu. ft. of water. They are measured by pouring water containing Daphnia into a tall graduate held in the sunlight. The individuals very soon settle to the bottom and the quantity may be determined with ease.

The water supply of the seven concrete propagating ponds

used is controlled by dams in such a way that a constant level is automatically maintained in each pond without overflow. The water is therefore stagnant at all times Each pond has an independent drain which leads directly into a bass-rearing pond. Hence all food organisms produced may be drained off directly into the rearing pond by removing a standpipe. During the last four years it has been customary to operate them in rotation, thus producing several crops in each during the summer season.

7

Shrimp Culture

Decapod crustaceans of the suborder Natantia (shrimps or prawns) are found in fresh and salt waters virtually all over the world and many of the larger species are highly valued as human food. In most countries, shrimp culture is nonexistent or at best in the experimental stages, but some species have been cultivated in southeast Asia for five centuries or more. The methods in use are crude, often consisting of no more than trapping and confinement of young shrimp in brackish water ponds for several months before harvesting. In most southeast Asian countries production of shrimp has long been incidental to the culture of brackish water fish, but mono-culture of shrimp, with concurrent technological advances, is emerging in a number of countries.

Truly intensive shrimp culture, with breeding in captivity, has yet to play a significant role in southeast Asia but has been part of the picture in Japan since 1934, when Motosaku Fujinaga achieved the first success in spawning and partial rearing of the kuruma shrimp *(Penaeus japonicus)*. Fujinaga strove to perfect his techniques until 1959 when, with the financial assistance of several fishery companies, he was able to set up a pilot hatchery and farm. By 1967, some 20 operators were using his techniques to produce 4000 tons of shrimp annually from 8500 ha of water .

There is presently no successful commercial shrimp culture in the Western hemisphere, but scientists and entrepreneurs in the United States and several Latin American countries have noted the similarity between American commercial shrimps and the kuruma shrimp and have attempted to apply Japanese culture methods, to date with only partial success. Attempts have also been made in the

United States to apply southeast Asian methods, but would-be shrimp culturists have been held back not only by biological and technological problems but by legal difficulties with regard to the ownership and use of estuarine water.

Shrimp of various species are fished virtually from pole to pole, but commercial shrimp culture is, with the notable exception of Japan, confined to the tropical and subtropical regions. In recent years, however, experimental shrimp culture has been initiated in temperate zone countries. Nor is freshwater exempt. Experiments in Malaysia and elsewhere have demonstrated the potential for culture of freshwater prawns.

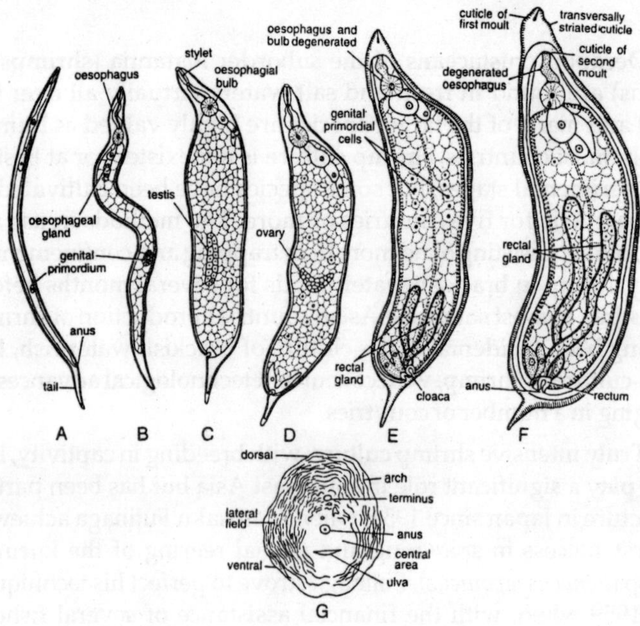

Fig. 7.1 Meloidogyne incogsita. A- Infective Larva, B-Second stage larva with genital primordium. C-Male Third stage Larva. D-Female. Third stage Larva, E-Third stage larva with second moult. F-Fourth stage female larva. G-Perineal pattern

The first step in site evaluation is sampling by cast net to determine the relative abundance of various shrimps. (The species list on p. 67 is in order of preference.) If large numbers of Caridean shrimps are captured, the site is likely to be rejected. There are seven additional factors important in site selection:

1. *Tidal range and swamp elevation.* Tidal inflow and outflow should ideally be such that the pond can be harvested every day, since premium prices are paid for very fresh shrimp. If the swamp is too high or high tides too low, no sea water will enter on neap tides, even at high tide, thus reducing the number of fishing days. In practice it is not possible to harvest any pond every day; culturists aim at 20 to 22 harvest days per month. Very low swamps and high tides provide more potential harvest days, but expenditures for construction and maintenance of dikes and sluice gates may be prohibitive under these circumstances.
2. *Depth.* It should be possible to maintain about 0.7 m. of water to the pond at all times.
3. *Location relative to the coast.* If the pond is too near the beach, the dike may be damaged by wave action. Usually a fringe of at least 15 m of mangroves is left undisturbed between the shore and the pond for protection. On the other hand, if the pond is too far from the coast it may be less accessible to shrimp and there may be a danger of flooding with rainwater during the monsoon season.
4. *Soil quality.* The bottom should be mostly hard clay, perhaps mixed with sand and organic detritus. Soft, porous bottom materials are to be avoided, since the swamp bottom is the source of construction material for the dike. There should be no more than 50 cm of silt on the bottom since, for unknown reasons, heavy silting reduces shrimp catches. Large amounts of organic matter are also to be avoided, since their decay may produce anaerobic conditions leading to mass mortalities of shrimp.
5. *Surface area.* The smallest ponds require at least two workers daily for maintenance and operation. For this reason ponds of less than 12 ha. are not economically feasible.
6. *Salinity.* Salinity of sea water entering a shrimp be at least 18% at all times, 24 to 30% is considered optimal.
7. *Drainage basin.* Heavy freshwater run-off can reduce salinity below the preferred level for shrimp. For this reason swamps which receive large freshwater streams

are not favored. If a swamp is otherwise suitable, however, it may be feasible to divert such streams.

Once a suitable swamp is located it is converted into a pond by surrounding it with an earthen dike, or bund. The bund must be high enough to prevent the highest tide from flowing over the top, strong enough to withstand water pressure differentials, and wide enough for operating personnel to walk on top. Construction is usually by manual labour. Given the soft soils of mangrove swamps, this is more efficient than dredges, tractors bulldozers, and so on Index where machinery has been used in bund construction it has been necessary to make the bund wider and higher merely to support the machinery.

Clay for bund construction is taken from the swamp bottom in slabs about 30 cm x 15 cm x 10 cm thick. These slabs are laid side by side around the perimeter of the pond and each layer is sun baked before the next is added. The quality of workmanship at this stage will largely determine the amount of maintenance necessary in the future. Even a well constructed bund will sink about 30 cm/year, but poorly constructed bunds may sink as much as 90 cm.

Clay slabs are excavated so as to create a pattern of channels radiating from the sluice gates to the farthest corners of the pond. Experience has shown that this practice increases yields. It may be that the channels serve to distribute young shrimp evenly so that they are less vulnerable to predators and make more efficient use of the food supply. Not only the channels but the entire bottom is cleared of mangrove stumps and levelled, since stumps and other irregularities provide hiding places for predators. The bottom should slope toward the sluice gates to permit total drainage if desired.

Sluice gates must be large and numerous enough to allow tides to enter and leave the pond rapidly, thus avoiding extreme pressures on either side of the bund. Sluice gates should be located so that the incoming tide will not generate eddies and whirlpools, which may trap the incoming shrimp. The gates themselves should be built with a concrete foundation and sides, and the bund adjacent to the gates should be reinforced by wooden pilings. A simple wooden windlass is used to operate the gates.

The channel connecting the pond to the sea should be straight or nearly so and kept clear of plants. It is believed that plants vibrating in the current frighten young shrimp.

Owing to the rapid turnover of water in such ponds, fertilization is not practiced. A few experimental attempts did not appear to produce any significant differences in yields. It is customary, however, to drain the pond completely at least once a year mid let it sun bake for 3 to 4 days before refilling This is believed to increase the rate of production of plankton and other microorganisms.

Next to normal sinking and erosion, the biggest problem in pond maintenance is the mud lobster *(Thalassina anomala),* which burrows in bunds, weakening them and causing leaks. Mudlobsters may be killed by placing a piece of calcium carbide about 25 mm in diameter into the funnel-shaped burrow. A small amount of quicklime poured into the funnel has the same effect.

Daily stocking of shrimp ponds is usually uncontrolled. When the incoming tide in front of the sluice gates reaches a height of 60 cm above the pond level, the gate is simply opened and shrimp and various other organisms are allowed to swim in. A 60 cm head is necessary to ensure that no shrimp swim out of the pond. At dead high tide a 0.13 cm mesh wire screen is placed across the gate to retain the young shrimp. Although there are undoubtedly seasonal spawning peaks for each of the cultured species, there is some reproduction of all species throughout the year, so that this method is practicable year-round.

This sort of stocking permits entry of numerous predators which, if not controlled, will substantially reduce the yield of shrimps. The most serious predators are fish, which may be effectively controlled by poisoning with teaseed cake every 4 months.

Ponds so stocked always contain shrimp at all stages of growth. Large shrimp, up to 4 cm in carapace length, are harvested at dusk or night on ebb tides as they seek to return to the sea. Daytime harvesting is not practiced since few of the shrimp are swimming about during the day. Shrimp are captured in a conical net about 9 m long with mesh decreasing from 2 cm at the mouth to 1 cm at the cod end. The mouth is constructed so as to fit in place of the sluice gate. The net is affixed directly behind the gate just before the tide ebbs. When the tide begins to ebb the gate is opened and shrimp attempting to leave the pond are captured. Damage to the net by the rushing current is averted by supporting it in a troughlike wooden raceway on the outer side of the sluice gate. When the water level in the pond is reduced to about 60 cm the sluice gate is closed and the net hauled up.

Reliable production figures are not available for this form of shrimp culture primarily because culturists are reluctant to reveal their incomes. Yields of 300 to 800 kg/ha are reported, but it is likely that well managed operations yield up to 1200 kg/ha of edible Penaeid shrimps annually.

As in all forms shrimp culture that depend on natural spawning and recruitment, the scheme envisioned by Knox depends largely on judicious site selection. The following factors appear to be crucial in growing shrimps:

1. *Soil quality*. The soil must be conducive to both the growth of shrimp and the construction of dikes. The latter factor of course implies little more than low porosity. Determining whether soil is chemically and physically compatible with shrimp is somewhat more complex. Complicating this decision are preferences of individual shrimp species. Brown shrimp and white shrimp generally prefer softer, muddier substrates than pink shrimp, which are commonly found on sandy bottoms. A more general requirement is the absence of metallic oxides, which are repellent, if not toxic, to shrimp.
2. *Pollution*. Domestic and industrial pollutants are of course detrimental to the welfare of shrimp, but a less obvious source of "pollution" is the mangrove tree. Although in Malaya shrimp ponds are built in mangrove swamps, the American species of man-grove at times impart to water a substance which is toxic to American shrimps.
3. *Salinity*. Postlarvae and juveniles of American Penaeid shrimps are tolerant of a wide range of salinities, but as they approach maturity it is important that excessive amounts of freshwater not enter the pond. The safe salinity range seems to be 22 to 37%.
4. *Depth*. Since high production of benthic organisms is desirable, 0.5 to 0.7 m is considered to be optimal, although depths up to 1.3 m may be satisfactory if the water is unusually clear. In general, deeper holes are detrimental in that shrimp will congregate there reather than near the sluice gate when the pond is drained for harvest. One or two deep holes of known location may, however, be advantageous in the Carolinas, where sudden chilling

may occur. Large-scale mortality of shrimp, particularly white shrimps, has been known to occur in shallow ponds when air temperatures drop to around 4 to 5°C.

5. *Land configuration*. This is primarily an economic consideration. Construction costs may be minimized by taking advantage of natural impoundments formed by tidal creeks, swamps, or lagoons which can be adapted for shrimp culture by merely placing a sluice gate across the mouth. Such sites in addition to being rare, are often too deep or, if located on a creek, are subject to desalinization in the wake of heavy rainfall. In the majority of instances, a dike will have to be constructed. The prospective culturist should thus seek to reduce construction costs by using existing land features such as river banks and railway embankments as part of the dike system and by constructing dikes using soil excavated at or near the pond site. All ponds should be separated from the open sea by at least 15 m of marsh grass or other vegetation to break storm waves.

FRESHWATER SHRIMP CULTURE

When shrimp are mentioned, one customarily thinks of a marine environment. But many of the largest and most desirable shrimps occur in freshwater. By far the best known of these is the giant freshwater prawn *(Macrobrachium rosenbergi)* of the Indo-Pacific region (Fig.). Maximum length of *M. rosenbergi* is about 25 cm for males and 15 cm for females. Both sexes bring extremely high prices relative to other seafoods.

Attempts have been made, with some success, to induce *M. rosenbergi* to enter impoundments for growing marketable size, but the rather low natural population densities characteristic of this species render commercial culture by the methods used in southeast Asia for marine shrimps less than practical. Accordingly, fishery workers in the Indo-Pacific countries have made efforts to spawn and rear *M. rosenbergi* in captivity. The first experimental success in this enterprise was achieved by an *FAO*, supported fisheries development project under the direction of S.W. Ling at the Fisheries Research Laboratory of the Fisheries Department of the Federation of Malaya at Glugor, Penang. Commercial culture there is not yet a reality but might well prove feasible with some improvement in the efficiency of the methods practiced at the Penang hatchery.

In nature, adult *M. rosenbergi* are found in virtually all types of fresh and brackish waters. Larval development, however, requires water of 8 to 22% salinity. Mating takes place a few hours after the female performs a premating moult in estuaries or in freshwater streams not far above tidewater. The male deposits sperm on the ventral thoracic legs of the female, but the sperm are not retained for long periods of time as in the marine Penaeids. Rather the entire complement of eggs is deposited and fertilized in brood chambers at the base of the thoracic legs within 6 to 20 hours after mating. Unmated females also deposit eggs after moulting, but these fall off after 2 or 3 days.

It is believed that, in nature, females may spawn 3 to 4 times a year, producing up to 120,000 eggs each time. The eggs remain attached to the female, who aerates them by beating her pleopods and carefully removes dead eggs and foreign matter, using the first pair of thoracic legs, up to the time of hatching. The process of incubation usually requires about 19 days at 26 to 28°C. From about the twelfth day the colour of the eggs, originally bright orange, begins to fade to a pale grey. When this darkens to slate grey, hatching is imminent.

From the first minutes of life, M. *rosenbergi* larvae are active swimmers, but they are not initially strong enough to hold their own against a current. Thus river-hatched larvae are quickly swept downstream to a suitably saline environment. At first the larvae swim together in large groups, but after the tenth day of life they tend to separate. They feed primarily on zooplankton, but in the absence of an adequate supply of live animal food they will take minute bits of dead organic material or plants. Within 35 to 55 days the larvae have passed through 12 stages and metamorphose into juveniles.

Juveniles immediately adopt a benthic mode of life and commence to feed on benthic animals and organic detritus. Moulting occurs every 4 to 6 days. Juvenile M. *rosenbergi* are believed to crawl slowly upstream until, 2 to 3 months later, in the recognizable form of a young prawn, many of them have reached pure freshwater.

At this time young *M. rosenbergi* are 5 to 6 cm long (measured from eye socket to tip of telson) and weigh about 6 g each. The young prawns continue the migration begun as juveniles but swim and crawl at a much more rapid rate. Some may travel over 60 km upstream from the sea. From this time on, M. *rosenbergi* will eat almost any piece of living or dead organic matter of suitable size. When sufficiently hungry they may resort to cannibalism. Under favourable conditions, sexual maturity is attained in 9 months, after which time there may be a downstream migration'.

Sometimes egg-bearing or "berried" female M. *rosenbergi* can be captured from the wild for use in culture. Three methods of capture are employed: trap, hook and line, and hand net.

Prawn traps are funnel-shaped structures of split bamboo or galvanized wire mesh. In use they are baited with pieces of fish, shrimp, or baked coconut and set overnight near the banks or rivers.

Hook and line fishing for prawns, using one or more barbless hooks baited with earthworms, small shrimp, or pieces of baked coconut, is practiced in both Malaysia and Thailand. A small sinker, just heavy enough to hold the baited hooks on the bottom, is attached about 25 cm above the uppermost hook. When a bite is detected a few seconds are allowed for the prawn to hook itself, then the line is drawn up steadily, without jerking. Berried female prawns caught on hook and line are kept for a day or so in submerged wire cages to ensure that they are not seriously injured.

Hand netting is done at night, when prawns move into shallow water to feed. A strong light temporarily stupefies the prawns and enables the fishermen to see the bright bluish reflection of their eyes. Specimens thus detected can be captured by careful netting from the rear.

Transportation of live prawns from the field to the culture site presents few problems. As long as they are kept moist, prawns will survive for several hours when packed in baskets between layers of soft plants. Longer journeys may require aerated tanks or plastic bags filled half-and-half with water and oxygen. If bags are used, the sharp tip of the rostrum must be cut off to prevent its puncturing bags.

If berried females are not available in sufficient quantities, young prawns may easily be reared to maturity in captivity. Mature females are usually kept in groups in aerated 100 to 200 litre tanks. Such groups must be constantly watched for moulting. Newly moulted females must be screened off from the rest, else they will be attacked while the shell is soft.

Males may fight at any time, so they are usually kept individually in 60 litre tanks. When a female has moulted, 2 to 3 hours are allowed for the shell to harden, then she is placed in the tank of a male. Mating will occur within a few hours, followed by egg laying within 24 hours. If sufficient freshly moulted females are available, group spawning may be practiced. For this purpose 2 to 4 males are placed together with 8 to 20 females in tank 2 to 3 m long, 1 to 1 ½ m wide, and 40 cm deep and mating and egg laying proceed as described. As in all stages of M. *rosenbergi* culture, vigorous aeration is essential. Both sexes are kept in freshwater through spawning.

Berried females, whether collected from the wild or mated in captivity, are transferred to individual 50 to 200 litre tanks. When the eggs begin to turn grey, sea water is added daily in small amounts so that by hatching time the salinity is 8 to 15%. This is not only better for the larvae than freshwater, it seems to produce better rates of hatching.

An average female produces about 50,000 larvae, which are almost immediately transferred to a rearing tank 2 to 3 m long and 50 to 70 cm wide containing 16 to 20 cm of water. The bottom of the rearing tank slopes slightly to a collecting pit at one end. Larvae are captured for transfer by shading all but one corner of the hatching tank. The larvae are attracted to the lighted corner, where many of them may be captured by dipping with a cup. The remainder are siphoned into the rearing tank. They are quite hardy with respect to water quality, but every effort is made to maximize growth and survival by providing near-optimal conditions; 12 to 14% salinity, a temperature of 26 to 28°C, and a *pH* of 7.0 to 8.0. Sudden changes in these environmental parameters are scrupulously avoided. If the salinity of the water in the hatching tank is less than 12%, the rearing tank is initially filled to less than its capacity with water of the same salinity as that in the hatching tank. Gradual upward adjustment of salinity is made by slowly siphoning a predetermined amount of sea water into the tank. An aerator and stirrer are used to keep the dissolved oxygen concentration near saturation.

The basic larva food used at Penang is brine shrimp nauplii, which are hatched in the rearing tank, using eggs imported from the United States. Brine shrimp eggs are placed within a floating ring at one end of the tank, which is shaded. The newly hatched nauplii are attracted to the lighted end of the tank and eaten. For the first few days of rearing the daily complement of brine shrimp eggs is about ¾ teaspoonfuls per batch of 50,000 larvae. This is gradually increased to 3 teaspoonfuls per day when the larvae are 30 days old, then held at this level until they metamorphose to juveniles. Some brine shrimp should be present in the rearing tank at all times to prevent cannibalism.

Table 7.1 Particle Size of an Artificial Feed Fed to Larvae of the Giant Freshwater Prawn (*Macrobrachium rosenbergi*).

Age at Larvae (Days)	*Food Particle Size (mm)*
2-4	0.4
5-10	0.8
11-20	0.9
20+	1.1

Since brine shrimp eggs are quite expensive, supplementary feeds are used as much as possible. A suitable natural food is fresh ripe fish eggs, thoroughly washed to eliminate ovarian tissue and other foreign matter. In 1968 workers at the Songkhla Marine Fisheries Station in Thailand, succeeded in rearing M. *rosenbergi* larvae using mullet (*Mugil* spp.) eggs as the primary food with brine shrimp nauplii as an occasional supplement.

Prepared foods are also used, although brine shrimp nauplii appear to be essential during the first day of life. The current favourite at Penang is a vitamin-fortified mixture of fish flesh and egg custard steamed together, drained and passed through a screen to provide particles of an appropriate size. Fish flesh or eggs custard may also be fed separately. Powdered dried chicken blood is successfully used in Thailand. M. *rosenbergi* larvae will also eat phytoplankton, and the Hawaii Division of Fish and Game has found it convenient to grow larvae in "green water" for the first 12 days.

Prepared foods are provided at the rate of approximately 30% of the total body weight of larvae per day. For convenience in feeding, most of the rearing tank may be shaded at feeding time to concentrate the larvae in one place. Feed is gently spread on the surface of the water with a medicine dropper. This should be done slowly enough that each larva has a chance to feed. If most of the larvae appear to be carrying food particles, they are being correctly fed.

Use of prepared feeds requires that great attention be paid to cleanliness. Uneaten food particles and fecal matter are siphoned off twice daily. Every 10 days or so a partial change of water is advisable. This is done by shading part of the tank and siphoning from that portion.

One of the disastrous consequences of overfeeding or lack of cleanliness may be an incurable fungus infection which manifests itself as small white patches on the tails and at the bases of the appendages of the larvae. Infected larvae should be removed and destroyed. If a large percentage of any batch of larvae are infected, it is best to sacrifice them all.

Larval M. *rosenbergi* are also subject to a serious protozoal infection which, if caught in the early stages, may be treated with 0.2 ppm of malachite green for 1/2 hour daily, or for 6 hours with a single dose of 0.4 ppm copper sulfate.

Prawn larvae are sensitive to nicotine, so tobacco smoking should be prohibited in the vicinity of larvae rearing tanks.

Yet another source of mortality among late larvae is jumping out of the water. Stranding may be averted by shading the sides of the tank to keep the larvae in the center.

When the larvae are ready to metamorphose, small branches, stones, shells, and so on, are placed on the bottom of the rearing tank to provide shelter for the freshly moulted juveniles, thus deterring cannibalism. When 90% of the larvae have metamorphosed, the remaining slow growers are transferred to another tank for further rearing.

The metamorphosed juveniles must next be accumulated to freshwater. This is accomplished within 3 to 8 hours by gradual replacement of the brackish water in the larvae rearing tank.

Juveniles may be reared in the same tanks used for the larvae, but it is better to provide more spacious quarters in the form of larger concrete tanks or earthen ponds with cemented brick walls. These facilities vary in surface area from 5 to 50 m2 and in depth from 15 cm to 1 m and are abundantly supplied with shelters for the young prawns. In addition to the usual aeration, it is desirable to maintain a slight flow of water through the tank or pond. Stocking rates for juveniles vary from 2 to 10/m2.

The juvenile prawns are very catholic in their food habits, but growth is maximized and cannibalism minimized by feeding fresh animal material as often as is economically feasible. Fresh fish, mollusks, and earthworms cut into pieces according to the size of the juveniles are the principal food, although whole live aquatic worms and chironomid larvae are used when available. Supplementary foods include dried animal material softened in freshwater for ½ hour before use and pieces of grains, peas, beans, and soft aquatic plants. Feeding is carried out four times daily; three daylight feedings and one at night. To avert cannibalism the amounts fed must be in excess of what the shrimp can consume. Insofar as possible all uneaten food is siphoned out after each feeding.

Juvenile M. *rosenbergi* are subject to the same diseases as larvae. As a preventive measure juvenile rearing ponds are completely dried, drained, and disinfected before and after each use. Disease treatment is the same as described for larvae.

With good management juveniles should grow to 2-to 3-cm prawns within about 30 days, with a survival rate of about 50%. When they reach about 4 cm in length the young prawns are suitable for stocking in production ponds. Almost any sort of pond over 200 m2 in surface area and 50 cm in depth, with a water temperature of 22 to 32°C, can he used for growing M. *rosenbergi*, but larger ponds, 1000 m2 or more in area and 1 to 1½ m deep, are more economical.

Ponds for prawn culture are prepared in the same general manner as fish ponds in southeast Asia: predators are eradicated, inlets and outlets screened, and aquatic plants removed. It is desirable to have a gentle flow of water through the pond at least a few hours each-day.

Prawns may be stocked alone or in combination with fish. One finds in the literature references to growing M. *rosenbergi* in combination with "carp." It should be emphasized that this refers to various of the Chinese carps, not to the common carp *(Cyprinus carpio)*, which competes for food with prawns and brings a far lower price. Fishes successfully used in culture with M. *rosenbergi* include such herbivores and/or plankton feeders as the big head *(Aristichthys nobilis)*, grass carp *(Ctenopharyngodon idellus)*, silver carp *(Hypophthalmichthys molitrix)*, catla *(Catla catla)*, rohu *(Labeo rohita)*, milkfish *(Chanos chanos)*, *Osteochilus hasselti*, *Barbus gonionotus*, grey mullet *(Mugil cephalus)*, kissing gourami *(Helostoma temmincki)* sepat siam *(Trichogaster pectoralis)*, and three-spot gourami *(Trichogaster trichopterus)*, Suitable stocking rates depend not only on the numbers and kinds of fish stocked, but on the quality of soil and water. Table 3 is a general guide to stocking practices.

Natural production within the pond supplies most of the food for prawns at this stage. Productivity is enhanced by monthly application of 200 kg of cow dung and 10 kg of lime per hectare. Supplementary feeding is also practiced, using 75% animal material, including small pieces of fish, mollusks, earthworms, offal, live insects and silkworm pupae, and 25% plant material, such as various grains and rotten fruit. Five percent of the total body weight of prawns is fed daily, half in the morning and half in the afternoon. Waste may be prevented by placing food in trays along the side of the pond.

Both food and shelter for the prawns may be provided by growing small patches of *Ipomoea* in the pond. The area covered by *Ipomoea* should not, however, exceed 10% of the area of the pond. Further shelter for moulting prawns may be supplied by placing branches on the bottom of the pond.

Table 7.2. Stocking Rates for Giant Freshwater Prawns *(Macrobrachium rosenbergi)* Cultured Alone and with Fish in Southeast Asia.

Pond Conditions	*Stocking Rate (Prawns/Ha)*	*Stocking Rate of Fish*
	Prawns Cultured Alone	
Rich	15,000	
Medium	10,000	
Poor	6,000	
	Prawns Cultured with Fish	
Rich	6,000	Full
	12,000	Half
Medium	4,000	Full
	8,000	Half
Poor	2,000	Full
	4,000	Half

The most serious management problem encountered in the phase of prawn culture is oxygen depletion, to which *M. rosenbergi* is more sensitive than most fishes. When prawns migrate toward the edges of the pond and appear sluggish in their movements it is time to apply remedial measures to increase the supply of oxygen.

Under favourable conditions prawns in ponds should reach marketable size (15 cm and 100 g) in 5 months, at which time they may be harvested by draining the pond or seining.

Less intensive culture of M. *rosenbergi* in paddy fields is also practiced, but the future of this form of prawn farming is threatened by the increased use of insecticides on rice. The bunds of paddy fields used for this purpose should be slightly raised so as to retain at least 12 cm of water throughout the 4 month growing period of the rice. Inlets and outlets should be equipped with screens to prevent the escape of prawns and entry of predators. These screens should extend about 0.3 m above the water surface or prawns will climb over them and escape. One or two small sump pits 1 m x 2 m x 50 cm deep should be constructed near the outlet to trap the prawns when the field is drained.

Paddy fields should be stocked only after the rice seedlings are fairly well rooted. Due to the shorter growing season, older prawns 2 to 3 cm longer than those used in pond stocking are preferred. These are stocked at about 1 prawn/15 m^2 under average conditions. Supplementary feeding is not practiced.

Since culture of M. *rosenbergi* is still largely in the experimental stages, there are few data on the yields which may be achieved. However, a commercial, albeit pilot, operation on Oahu, Hawaii, under the supervision of the State's Department of Fish and Game can produce 3,000 kg/ha of giant prawns. Larval survival rates in a highly automated hatchery are satisfactory even though they could still be improved. At moderate flow-rates of water through the adult rearing ponds, limits on stocking are mostly conditioned by dangers of disease. But even lacking experimentation with a more vigorous flow, T. Fujimura, who has pioneered this success, is confident that 4,000 kg/ha. are attainable with continuous stocking and harvesting schedules.

Large freshwater shrimps similar to M. *rosenbergi* are found in practically all tropical and subtropical regions, and *Macrobrachium rude* is coincidentally cultured along with M. *rosenbergi* in India, but the only other freshwater species thus far studied to any extent are *Macrobrachium malcolmsoni* and *Macrobrachium carcinus*. In Pakistan, the Directorate of Fisheries has succeeded in experimentally breeding and rearing M. *malcolmsoni* in a freshwater pond (size not given). Stocking with about 15,000 young yielded 560 kg of adult shrimp. Breeding was accomplished by confining pairs in very fine-meshed cages, 68.5 cm x 35.5 cm x 18 cm deep at the side of the pond and feeding with rice bran each morning.

Macrobrachium carcinus has been experimentally reared through all larval stages at McGill University's field station in Barbados, but mortality has been too high for effective culture. As far as is known, neither spawning nor hatching of eggs in captivity has been accomplished.

It is reported that in Peru larvae of the freshwater shrimp *Macrobrachium caementarius* are harvested from the lower end of rivers, raised in tanks for a few months, and used to stock both private and public waters.

Prospectus

Further growth of the shrimp culture industry will result from expansion into new parts of the world, improvements and innovations in culture techniques, and introduction to culture of new species. Most of the countries which presently support commercial or experimental shrimp culture have been mentioned here. New additions to the list may be expected regularly, as virtually every country with a seacoast has some potential for shrimp culture.

New Techniques

A number of new and/ or experimental techniques have also been mentioned here. At present the chief stumbling block in shrimp culture is poor survival of the larvae. The key to better survival is apparently a better diet. Japanese culturists have achieved the greatest success in supplying the nutritional needs of young shrimp, but their reliance on imported brine shrimp is costly and, if imitated by many other countries, might result in the depletion of brine shrimp stocks in the United States. Mention has already been made of the successful substitution in Thailand of fish eggs for brine shrimp in the diet of *Macrobrachium rosenbergi*. It has been suggested that amphipods or other locally available small crustaceans might be cultured as food for young shrimp in countries where brine shrimp are not available.

Improvement might also be effected in the feeding procedures for older shrimp. It would be particularly advantageous to culturists if a prepared food comparable to the pelleted feeds used in culture of carp, trout, and other fishes could be developed for shrimp. A suitable food pellet would have to incorporate a superior binding agent, since shrimp, rather than ingesting pellets whole as do most fish, hold them and pick them apart. In a British study *Palaemon serratus* were fed with pellets containing one of eight different protein foods. In each case the prepared feed produced much poorer growth than fresh food containing the same protein source. When, instead of a protein source, powdered polyethylene was added to the basic formula as a "nutritionally inert" filler, growth was better than that achieved using five of the other eight experimental feeds. Experimental work on pelletized diets for shrimp was also carried out at Florida State University, where a food conversion ratio of 3:1 in pink shrimp under experimental conditions was reported.

Other promising experimental techniques include growing shrimp in thermal effluent and monosex culture. The former method is being studied chiefly in the United States where numerous large power plants release large quantities of heated water used for cooling purposes. Conclusive results are not yet available, but indications are that growth and survival of Penaeid shrimps may be enhanced in "thermally enriched" environments.

Monosex culture has not yet been attempted, but its feasibility is suggested by the fact that female Penaeid shrimps of most, if not all species are larger than males of the same age; female sugpo prawns cultured in the Philippines average two to three times as large as the males at harvest. Research in West Germany has disclosed that female sand shrimp convert food more efficiently than males. On the other hand, male *Macrobrachium rosenbergi* grow much larger than females. It seems likely that similar sexual dimorphisms exist in other groups of shrimp as well.

New Species for Culture

Nearly all attempts atr shrimp culture to date have involved Penaeid shrimps or members of the palaemonid genus *Macrobrachium.* Only a minority of these species have been cultured commercia'ly and few of the remainder have been adequately studied, so additional members of these groups may be expected to enter the picture. Of particular interest among the Penaeids are those exceptional species which are highly euryhaline. Notable among them are the Australian greentail prawn *(Metapenaeus benettae)* and *Metapenaeus ensis,* both of which spawn naturally in estuaries, a trait that could greatly simplify estuarine culture. Post-larval greentail prawns apparently prefer water of less than 20% salinity. Australian biologists are currently experimenting with the culture of these and/or similar species, locally called "greasy backs."

Edible non-Penaeid shrimps which possess the ability to reproduce in brackish water include *Caridina gracilirostris,* most species of *Macrobrachium, Palaemon,* and *Leander,* and some species of *Palaemonetes. Macrobrachium lanchesteri* of Malaysia will even reproduce in stagnant freshwater. In addition to possessing the considerable advantage of being very hardy with respect to high temperatures, low dissolved oxygen concentrations, and waters with low mineral content, M. *lanchesteri* is believed to be herbivorous.

In concentrating on members of the genera *Penaeus, Metapenaeus,* and *Macrobrachium,* shrimp culturists have overlooked species with less complex life cycles. Such species, if they could be cultured, would eliminate many of the difficulties now encountered in rearing young shrimp. These species may be roughly divided into four categories :

1. *Species with no free-swimming larvae.* In *Sclerocrangon boreas* of the North Atlantic and Pacific and *Sclerocrangon ferox* of the arctic Atlantic the young remain attached to the female's pleopods and do not feed up to the zoea stage. These two benthic species reach lengths of over 12 cm and appear potentially suitable for culture. Certain other members of their genus, however, produce pelagic larvae.

   There are species of shrimp which bypass the larval stages entirely and emerge from the egg as juveniles, but most of them are too small for human consumption. An exception is *Bythocaris leucopia,* which exceeds 9 cm in length. *B. leucopia* is a deep-water species, however, and thus might not be amenable to culture. A better possibility might be the Japanese *Pandalopsis coccinata.* The 15-mm shrimp which emerges from the egg of this species is classified as a zoea, but is benthic and probably has feeding habits similar to the adult.

2. *Species with larvae which do not feed.* There are species of shrimp which have no functional mouth parts as zoea

and subsist on yolk until they become juveniles. Most of these are deep water pelagic species and would probably be difficult to maintain in captivity. A possible exception is the benthic *Glyphocrangon spinicauda.*

3. *Species with large larvae.* Much of the difficulty in rearing shrimp larvae stems from their small size. Larger larvae could be fed such readily available animals as copepods or amphipods or might even take prepared foods. Large larvae, however, mean large eggs, hence these species produce less eggs than the commonly cultured shrimps. Extreme cases such as *Richardina spinicincta,* which may produce as few as six eggs, are obviously unsuitable for culture, but species which produce several hundred eggs might prove satisfactory if extremely high survival of larvae were achieved.

4. *Species with few larval stages. Panadalus kessleri,* which is fished, although not cultured, in northern Japan, hatches as a very advanced 8-mm zoea and moults to become a postlarva within a few days. *P. kessleri* has the disadvantage of being very stenohaline, thus unsuitable for tidewater culture. The same advantages and disadvantages apply to *Pandalus platyceros,* which is currently being cultured experimentally in the United Kingdom. Other species that appear to have similar rapid development include *Argis lar* and perhaps other members of that genus; the holarctic *Lebbeus polaris,* and *Lebbeus groenlandicus,* found from northeast Asia to Greenland.

The species just mentioned represent only a tiny fraction of the known species of Natantia. Many of the rest are obviously unsuited for culture by virtue of size, habits, or habitat. Nevertheless, there are undoubtedly others of potential value to the culturist. It should be noted that almost all of the species discussed are boreal or tropical in distribution. It is thus particularly likely that additional cultivable species will be discovered in the South Temperate and Antarctic regions.

Whatever species or techniques are used in shrimp culture, it is unlikely that dramatic improvements will drive the price of shrimp down. The generally carnivorous habits of shrimp, along with the great amounts of energy lost in moulting, virtually guarantee that food conversion will be inefficient. Thus as long as there are ample markets for luxury foods the profit potential of shrimp culture will remain high. But if worldwide or, in some instances, local food shortages develop, shrimp culture may be overshadowed by biologically and economically more efficient forms of brackish water aquaculture.

8

Lobster Culture

One of the most highly prized of all seafoods is lobster, and efforts have been made to culture this delicacy at least since the 1860s. From the biologist's or culturist's point of view "lobster" may indicate either of two very different types of animal. Seafood gourmets find considerable difference between the two as well, but many consumers are less discriminating. The animals most of us visualize when lobster is mentioned are the American lobster *(Homarus americanus)* and its European counterpart, *Homarus vulgaris,* both of which are restricted to the north Atlantic. However, much of the lobster tail of commerce (including that marketed as rock lobster, Australian lobster, or South African lobster) comes from one or another of the spiny lobsters, which are found virtually throughout the oceans. There are many species of spiny lobsters belonging to several families, but all may be distinguished from *Homarus* by the lack of the formidable claws or chelipeds. A more important difference to the culturist is the much more complex larval development of spiny lobsters, which has kept spiny lobster culture lagging behind that of *Homarus,* although neither animal has yet been reared commercially.

Hatching and Rearing the Young in Captivity

Early lobster hatcheries relied on eggs taken from "berried" females and hatched separately with the aid of a device to agitate the water. It is currently believed that the female lobster is a more effective incubator than any man-made device. Either way hatching lobster larvae has always been easier than rearing them and, although first-stage larvae were stocked along the Atlantic coast of the United

States and Canada, it has been the opinion of most lobster, culturists that for a stocking program to be effective the young lobsters must be capable of assuming a benthic life upon release. Thus rearing of larvae has long been a preoccupation of lobster culturists.

The largest lobster hatching and rearing operation today is that carried on by the state of Massachusetts on the island of Martha's Vineyard, under the direction of John T. Hughes. Hughes and his associates have succeeded in mating lobsters in captivity, but the chief source of eggs for hatching continues to be the lobster fishery. Spring-caught females bearing brown eggs ("Brown eggers") are selected, since their eggs will hatch in a few months, whereas "green eggers" would have to be held over winter. The prospective mothers are placed in running water (salinity 30 to 31%) in hatching tanks 27.4 cm x 91.5 cm x 30 cm deep, divided into compartments for individually holding up to 70 lobsters. (Berried *Homarus vulgaris* reportedly cannot be kept together because they will claw the eggs off each other, but this is not so with *Homarus american us)*. They are fed on shell fish viscera and/or fish. Hatching usually commences in mid-May, when the water temperature reaches 15°C, and becomes intensive during a 2-week period in June and July.

The newly hatched larvae are swept away by the current and collected in a wire screen box then transferred to circular fiberglass tanks with a diameter of 400 mm and concave bottoms. Experienced workers estimate the number of larvae by eye and place about 3000 in each tank. Water is circulated through the rearing tanks to keep the larvae drifting in mid-water. If they were to accumulate on the bottom, cannibalism would severely decimate their numbers. Water enters the tank from the bottom through an inverted, perforated plastic cup, which breaks up the flow and keeps water circulating evenly in all parts of the tank. The outlet is an overflow pipe equipped with a similar device.

The first fry feed used was finely ground beef liver, but ground clams and frozen adult brine shrimp *(Artemia)* have been found to produce much better survival. Crab viscera are another acceptable food. Ground clams or brine shrimp are mixed with sea water in a proportion determined by the demands of the larvae, and one tablespoon of the mixture fed every three hours throughout the larvae's stay in the hatchery. Heavy feeding helps avert cannibalism, but overfeeding may clog the inlet and outlet pipes, thus reducing circulation and increasing cannibalism as well as raising the threat of pollution.

Using these methods an average of 22% survival to the third moult has been attained, with a record season's survival of 42.6%. Next to cannibalism the major cause of mortality is a gas disease caused by supersaturation of the water with nitrogen, a condition brought on by air leaks in the circulating system. Lobster larvae are sensitive to certain metal ions, but this has become less of a problem since lead piping was replaced by plastic.

The lobster fry in hatcheries are attacked by a parasitic suctorian *(Euphelota* sp.). Heavy infestations cause mass mortality by offsetting the larvae's buoyancy and forcing them to the bottom. Chemical treatment has proved futile, but *Euphelota* can apparently be excluded from hatcheries by filtering sea water through sand before it is used.

CULTURE OF PALINURID LOBSTERS

There are 7 genera and about 30 species of spiny lobsters (superfamily Scyllaridea), of which at least 15, all members of the family Palinuridae, are commercially important. So far as is known, all the commercially important species have similar but very complex life histories.

Table 8.1 The commercially important species of spiny lobsters and their ranges

Species	*Range*
Jasus lalandii	Australia, Juan Fernandez Islands (off coast of Chile), New Zealand, South Africa, Tasmania, Tristan da Cunha.
Jasus verreaxuxi	Southern Australia, New Zealand, Tasmania
Palinurus elephas	Great Britain to the Mediterranean
Panulirus argus	Florida, the Caribbean area, the Atlantic coast of South America, Bermuda
panulirus gracilis	Southern Mexico to Ecuador
Panulirus inflatus	Gulf of California to Southern Mexico

Panulirus interruplus	Southern California, Lower California (Mexico)
Panulirus japonicus	Japan
Panulirus laevicauda	Bermuda to northeastern South America
Panulirus longipes	Western Australia, Ryukyu Islands, Taiwan, Philippines, New Caledonia
Panulirus marginatus	Hawaiian Islands
Panulirus ornatus	Indian Ocean, Ryukyu, Islands, Taiwan, Philippines
Panulirus penicillatus	Indo-Pacific, Korea, Lower California (Mexico), Galapagos, Costa Rica
Panulirus regius	Western Mediterranean, West Africa
Panulirus versicolour	Indo-Pacific

Mating occurs in inshore waters at various times of year, depending on species and age of females. Old females mate earlier than young and some of them are thought to mate twice a year. In mating males extrude sperm in the form of a viscous fluid which attaches to the underside of the female. There the surface rapidly hardens to form a sperm sac. Some time later, the eggs are extruded and pass over the sperm sac. At that time the female breaks the sperm sac by scratching it with her legs, thus fertilizing the eggs. They are then passed on to the swimmerets where they are attached, to remain until hatching. The number of eggs produced by a female varies with size and species. Estimates as low as 50,000 and as high as 4,000,000 have been recorded. The time between mating and fertilization is probably dependent on water temperature. In general it appears to be less than that for *Homarus*.

Certainly hatching occurs much sooner than in *Homarus;* only 3 weeks are required in some species. During this time the eggs change colour from bright red-orange to a dark brown which fades until they are almost colourless just before hatching.

If, from the culturist's point of view, the reproductive process of spiny lobsters is more efficient than that of *Homarus,* the spiny lobsters more than make up for it with protracted and complex larval development. The newly hatched larva is a flat, leaflike, and unusually delicate animal 2 to 3 mm long, known as a phyllosoma.

Phyllosomas are planktonic and float horizontally with the legs extended. At least some species are negatively phototropic and make fairly extensive diurnal vertical migrations. The larvae remain planktonic while passing through a large number of moults, most of which do not result in metamorphosis, although progressive changes in form may be detected. Usually 3 to 6 months and 6 or more metamorphoses are required to reach the puerulus stage, but in one laboratory experiment on *Panulirus japonicus,* 16 moults were passed through in 178 days without metamorphosis. It is possible that the number of larval moults is not fixed for a species and that frequency of metamorphosis is partially dependent on nutrition or some such factor.

The puerulus is superficially similar to the adult but is transparent, lacks lime in the skeleton, and may still be planktonic. Finally, the puerulus moults to become a juvenile about 2.1 cm long and settles to the bottom. After two or three more moults it acquires the reddish-brown colour of adult spiny lobsters. Survival to this point is extremely low since predation is extensive at all stages.

Juvenile and adult spiny lobsters are also subject to predation but avoid it to some extent by hiding under rocks and other cover during the day and foraging at night. Muddy bottoms and strong currents are avoided. In nature adult spiny lobsters consume a wide variety of foods, including fish, worms, mollusks, and smaller crustaceans. Cannibalism may occur if foods containing calcium carbonate are not available. Spiny lobsters may act as scavengers but exhibit a marked preference for fresh food.

Frequency of moulting, and therefore rate of growth, varies with food supply, water temperature, and sex. Mature females moult but once or twice a year, before mating and sometimes after hatching or otherwise disposing of their eggs. Males grow larger than females and may moult at any time of year, thus are thought to moult more often. Growth between moults is about 5 to 10% of body length for *Panulirus argus.* Maximum size and age varies with species. *Panulirus interruptus* is said to reach a weight of over 13 kg, but such specimens are rare, and 3 kg would be considered large. In California this species is believed to take at least 7 to 9 years to reach the legal minimum size of 10½ in. (27 cm), although *P. argus* may attain this length in 3 years.

Attempts at Culture

Obvious difficulties notwithstanding, many attempts have been made to rear spiny lobsters from the egg in captivity. Perhaps the first efforts were those of the California Department of Fish and Game, starting in 1911, with *Panulirus interruptus*. Sporadic efforts with that species have continued, but little progress has been made, though Margaret Knight at Scripps Institute of Oceanography in La Jolla, California, has been able to hold larvae in finger bowls for up to 4 months. Soviet scientists also report some success in holding and rearing *Palinurus elephas* in aquaria and nursery ponds.

Other members of the family Palinuridae which have been hatched and held in the laboratory for 5 to 10 moults include *Panulirus japonicus* in Japan, *Panulirus argus* in Florida, and *Jasus lalandii* in South Africa. Mention should be made of the confusion between the terms "stage" and "moult." There are descriptions in the literature of rearing *Panulirus japonicus* and other species as far as the "tenth stage," but when compared to natural larvae of *Panulirus interruptus*, for which the larval stages have been thoroughly described, the supposed tenth-stage larvae appear to be third or fourth stage. This confusion is occasioned by the lack of any direct correspondence between developmental stage and number of moults. Under unfavourable conditions of temperature or food supply, spiny lobster larvae may moult without appreciable growth or morphological change. The term "stage" should be used only to refer to morphologically distinct forms.

Other Palinurid species which have been hatched in captivity but not reared beyond the very early stages include *Panulirus inflatus* in California, *Panulirus longipes* in Japan, and *Panulirus polyphagus* in Malaysia. The necessity of unpolluted water free from silt or small bits of detritus on which the larvae may entagle their long legs and the need for nearly constant temperature would certainly constitute serious restraints to commercial culture, but the main problem at present is feeding. Early larvae of most species accept brine shrimp nauplii quite readily, but attempts to feed older larvae on adult brine shrimp, sea urchin eggs, and larval gobiid fishes have met with slight success. Complicating the situation is the fact that almost nothing is known of the natural food of spiny lobster larvae.

CULTURE OF SCYLLARID LOBSTERS

All of the commercially important spiny lobsters belong to the family Palinuridae. The related Scyllaridae are scarcely exploited. However, it should be pointed out that those species whose habits are most conducive to fishery exploitation are not necessarily those best suited for culture. Some Scyllarid lobsters possess the advantage for culture of having shorter and less complex larval development. In general they also appear to be hardier animals.

The Scyllarids *Ibacus ciliatus, Ibacus novemdentatus, Parribacus antarcticus,* and *Scyllarus bicuspidatus* have been hatched in Japan but have not been reared for more than a few days. Similar lack of success has been experienced in India with *Scyllarus sordidus.* However, Phillip B. Robertson of the University of Miami's Institute of Marine and Atmospheric Sciences has become the first to rear any spiny lobster to the juvenile stage. He reared the sand lobster *(Scyllarus americanus)* from the egg through 6 to 8 phyllosoma stages to metamorphosis in 32 to 40 days at 25°C and salinities of 23.2 to 38.6%. Most important from the culturist's point of view, Robertson was able to achieve this on an exclusive diet of brine shrimp nauplii. Robertson's success notwithstanding it will almost certainly be a long time before larval culture of the sand lobster or any other scyllarid lobster can be recommended for other than experimental purposes.

REARING OF SPINY LOBSTER PUERULI

Clearly, if there is to be commercial culture of Palinurid or Scyllarid lobsters in the near future it will involve capturing pueruli or juveniles and growing them in confinement. This possibility has been considered in Australia where young *Panulirus longipes* can be captured in abundance on the west coast. To retain these young for culture would presently be illegal, but if it could be shown that culture was economically feasible and would not deplete natural stocks, the legal restraints might be removed. Most undersize *P. longipes* taken off Australia are 2 to 3 years old and would need to be held for only 1 to 2 years to reach marketable size. Experiments have shown that the food conversion ratio of such animals fed on fish or ablone is about 6 : 1, which is quite efficient for a crustacean. However, taking into consideration the costs of food, capture of stock, pond construction, maintenance, and so forth, it seems unlikely that such culture would be profitable in Australia at this time.

The type of culture suggested for Australia has reportedly been attempted with *Panulirus japonicus* by Japanese researchers. Young *P. japonicus* caught in traps have been held in ponds or shallow bays and fed trash fish. The results indicate that it may be feasible to produce salable lobsters at a profit.

There may be other countries where the practice of capturing and growing young spiny lobsters will prove profitable. The feasibility of such culture would be enhanced if growth rates could be increased as has been done with *Homarus americanus*. This might be done by improving the quality of food fed to spiny lobsters. Most Palinurids in captivity are fed fish, probably for reasons of availability, but crustaceans have been shown to produce better results. A better possibility for increasing growth rates is the use of heated water. Steven A. Serfling of San Diego State College in California has been able to increase the growth rate of young *Panulirus interruptus* in the laboratory by as much as 260% over average natural rates by maintaining them at 28°C. It is not known whether this would be feasible within the framework of commercial culture.

9

Culture of Crayfish

Among the best known and most highly esteemed crustaceans are the lobsters *(Homarus* spp.), but they are rivalled as a delicacy by their smaller freshwater counterparts, the crayfishes or crawfishes. (Not to be confused with the marine spiny lobsters, family Palinuridae, sometimes marketed as crayfish or crawfish). True crayfish comprise more than 300 species and are found on all the continents except Africa. Although crayfish are esteemed as a gourmet food in several European countries and are the primary source of protein for certain tribes in New Guinea, they are generally underutilized by man.

Crayfish have attained importance as a commercial food product in parts of Europe and the United States. This most enthusiastic consumers of crayfish are the French, and crayfish farms have been in operation in France since 1880. But the most important crayfish producing area is Louisiana, the only American state with a history of French culture. In a good year more than 800,000 kg of "wild" crayfish may be caught and marketed in Louisiana; during bad years the yield may be less than half that figure. Fishery production is supplemented annually by 1.2 million kg produced on the 6000 to 7000 ha. of crayfish, farms in the state. Details are not available on techniques of crayfish culture in Europe, hence this report concentrates on practices in Louisiana.

Natural History of the Principal Cultured Species

There are twenty-nine species of crayfish known to inhabit the waters of Louisiana, but only two are cultured. On most farms the dominant species is the red crayfish *(Procambarus clarkii),* but a few areas produce chiefly white crayfish *(Procambarus blandingi)*. The natural history of the two species is similar.

Mating occurs in open water in the late spring, when the water level in the Louisiana swamps is high. At that time the male crayfish deposits sperm in an external receptacle on the female.

Shortly after the peak of the breeding season, female crayfish come out on shore and dig burrows near the water's edge. Since they are exposed to terrestrial predators at this time, areas well protected by emergent plants are preferred. Burrows are essentially vertical and usually 0.7 to 1.0 m deep, unless the water table is unusually high, in which case burrows half that depth may be found. By the end of July, all adult females have constructed burrows, which are then usually occupied by a single male, along with the female. Each burrow is capped with a plug of dirt. Young crayfish and unpaired males also seek shelter in the mud or in naturally occurring holes during the summer, but they do not burrow. Some additional mating may occur in the burrows, but the crayfish are thought to be essentially inactive until September, when egg laying occurs.

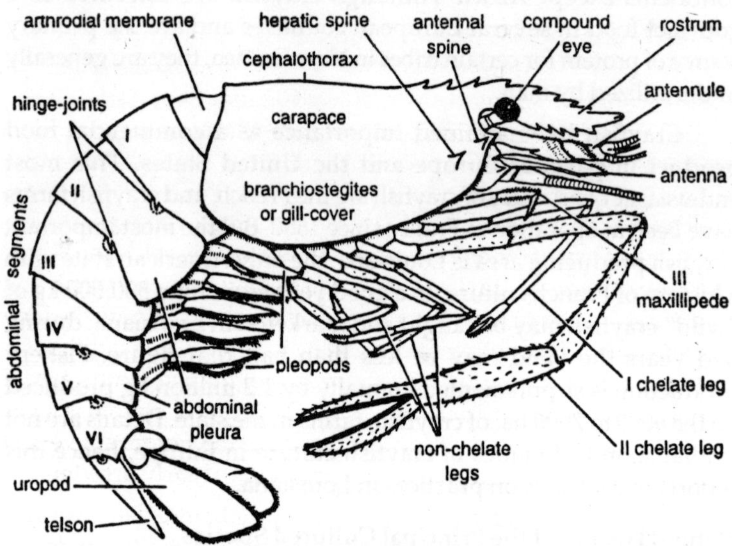

Fig. 9.1 Palaemon.

Eggs and sperm are simultaneously released by the female and the fertilized eggs attached to the underside of her tail until hatching, which occurs in 14 to 21 days with red crayfish and 17 to 29 days with white crayfish. The adults die soon after hatching. An average hatch is about 400 red crayfish (maximum 700) and somewhat less for white crayfish. Fertilization and egg laying often occur in the burrow, but growth and survival of the 25-mm, free-swimming young is greatly enhanced if open water is available. In years when the water is low in the fall, hatching may occur in the burrow, in which case the young suffer from overcrowding and lack of food. Or the female may crawl off overland in search of water and die of dehydration or be captured by a predator.

Young crayfish usually seek shelter in dense plant growth near shore, then move out to deeper water as mature. Both red and white crayfish are annual animals, and maturity may be reached in less than 6 months after hatching.

Crabnasdyfish of all ages are omnivorous but prefer animal matter. Nevertheless, most crayfish, both in nature and in culture, must subsist on a predominantly vegetable diet.

Pond Culture in Louisiana

Crayfish culture is practiced in two types of water rice fields and artificial impoundments. The latter, though its large-scale use dates back little more than 20 years, now accounts for more than 80% of the area devoted to crayfish farming.

Site Selection and Construction

As in most other forms of aquaculture, site selection is crucial in pond culture of crayfish. Due to the shallowness of crayfish ponds, it is essential that the land chosen be relatively flat. When the highest point in the pond bottom is covered by 0.3 m of water, depths of over 1 m should not occur in more than 25% of the remaining area. It is not, however, of any particular value if the pond bottom is level. In fact, small elevations in the bottom have the favourable effect of increasing the area available for burrowing of early breeders.

Densely wooded areas are not favoured, since dense shrubbery and trees along pond banks hinder harvesting, shade out desirable aquatic plants, add leaves and debris which decay and reduce oxygen levels, and may keep temperature in the shallows below optimum. It was in such environments, however, that pond culture of crayfish had its start in Louisiana and 2400 ha. of densely wooded swampland are still in use in an area just west of the Atchafalaya Basin, where conventional pond construction would be extremely difficult. These leveed swamplands are also heavily used by waterfowl, which are extensively hunted in Louisiana.

Soil quality surely has some effect on crayfish production but, in Louisiana at least, the primary consideration is that the soil hold water.

Water supply is of course important, but crayfish are very tolerant of naturally occurring physical and chemical conditions. Aspects of water quality which are of particular importance to the crayfish farmer are temperature and hardness. Optimum temperatures for red crayfish are thought to be 21 to 29°C, but growth is not drastically reduced until the temperature falls below 13°C. Above 32°C, red crayfish burrow into the mud and become inactive. White crayfish do better at slightly lower temperatures and begin to burrow at temperature as low as 27°C. Growth is the primary reason for the culturist to be concerned with water temperature, since neither species is likely to suffer significant thermally induced mortality at temperatures common in Louisiana. For the same reason, dissolved oxygen concentration, though unlikely to be a cause of mortality, should be kept as high as possible. .

Cultured crayfish have been observed to do well at pH levels as low as 5.8 and as high as 8.2. Crayfish in acid waters tend to have thinner shells. Soft water also results in thin, soft shells as well as poor growth and survival. It appears that the water in crayfish ponds should have a total hardness of at least 50 ppm, and that up to 200 ppm is desirable.

In recent years interest in crayfish farming in the coastal swamps of southern Louisiana has increased, and some attention has been paid to the effect of salinity on crayfish. Preliminary experiments indicate that crayfish will reproduce and grow fairly well at salinities of 6 to 10%.

Tolerant as crayfish are to natural conditions, they are extremely sensitive to synthetic chemicals and other substances which may find their way into agricultural water supplies. Such commonly used compounds as Pyrethrum, creosote, orthodichlorobenzene, sodium cyanide, turpentine, orthocresole, cresylic acid, pine oil, nicotine, carbon bisulfide, phenothiazine, calcium cyanamid, and chlorinated hydrocarbon pesticides have all been shown to be toxic to crayfish. The last-named group presents particular problems because of the possible danger to human health from dosages sublethal for crayfish and because the crayfish farmer may find it difficult to prevent contamination of his water supply by pesticides used on nearby agricultural land. It is recommended that crayfish culturists located in the vicinity of intensive terrestrial farming operations alert neighbouring farmers and crop-dusting pilots of the location of their crayfish ponds and attempt to secure their cooperation in averting contamination. When spraying is going on nearby, pumping of water from streams and ditches into crayfish ponds should be suspended.

The danger of contamination by pesticides and other pollutants is greatly reduced if the water source is a deep well, but many farmers, for reasons of necessity or economy, use surface water. Whichever type of water is used, it is best if each unit of the farm is provided with its own pump. Units need not be discrete but may be portions of larger ponds separated by inside levees. The smaller the units, the greater the ease of water quality control and general management. On the other hand, the more small units are built, the greater the construction cost per hectare of water surface. A large farm might consist of one or two 100-to 300-ha ponds broken up into 10 or so units. One-family farms as small as 8 ha exist but do not provide a full family income.

Detailed advice on levee construction may be obtained from the Soil Conservation Service, county agricultural agents, or commercial contractors. In general, levees should be high enough to keep out flood waters and wide enough to permit access to vehicles.

Water circulation has been found to be a deciding factor in the success or failure of many crayfish farms. To facilitate circulation it is necessary to provide not only an adequate inflow of water but good drainage as well. Drain pipes, which must be screened with 12-mm or smaller wire mesh, should be located as far away from pipes as possible to permit thorough mixing of water. Drains should be large enough to permit complete draining in 30 days.

Stocking

Although some ponds contain natural populations of crayfish, stocking is nevertheless necessary the first year, after which the population should be self-sustaining. The usual procedure is to stock ponds with adults in May to July, at which time crayfish are "tough" and prices low. The pond should be flooded for at least 2 weeks before stocking and arrangements made to purchase freshly caught crayfish. Stock held in captivity for even so short a period as overnight should not be accepted. Large specimens, about 30 to 45/kg. are preferable. Suggested stocking rates for red crayfish vary according to the amount of cover in the pond.

At least 80% of most lots of crayfish purchased for stocking in most parts of Louisiana will be red crayfish. If as occasionally happens, a stock of predominantly white crayfish is obtained, the figures in Table 9.1 should be increased by 25%.

Crayfish of either species should be placed in the pond as soon as possible and kept cool and damp until that time. Predation on newly stocked crayfish will be reduced if they are released in densely vegetated areas or, if the pond is sparsely vegetated, in deep water far from shore.

Table 9.1 Suggested Stocking Rates for Crayfish in Different Types of Ponds in Louisiana

Wild Crayfish Present	*Type of Pond*	*Stocking Rate (Kg/Ha)*
Yes	—	20-25
No	Densely vegetated	40-45
No	Wooded	45-60
No	Open, sparse cover	45-60
No	Open, very little cover	60-100

J.G. Broom at Auburn University obtained experimental evidence that stocking young red crayfish in March at 42,500 to 85,000/ha. would produce better yields than the conventional stocking system, but as far as is known this method has not been put into practice, perhaps due to the difficulty of capturing young crayfish for stocking. Broom suggests the following stocking formula:

$$\frac{\text{desired yield (kg/ha)}}{\text{weight at harvest}} \times \frac{100}{\text{expected survival}}$$

$\times$ number of ha. to be stocked = number of crayfish to stock

Manipulation of the Water Level

The most important management techniques in crayfish culture involve manipulation of water level and quality. Each year ponds are drained in late June or early July, when the females have started burrowing. Slow draining is preferable, for fast draining will strand some crayfish which are not ready to burrow and expose them to predators. Similarly, slow draining allows young crayfish to seek hiding places for the summer.

Ponds are reflooded in September to ensure that the newly hatched young will have ample water. If growth rates are normal, they should be ready for harvest starting in late November or early December. The culturist should keep on an eye rainfall, which is the principal controlling factor on wild crayfish harvests, and time reflooding so that he can begin harvesting before the wild crop comes in. Early crayfish bring the best prices and may spell the difference between economic success and failure.

Ordinarily, harvesting lasts until the pond is drained the following summer. Between reflooding and the onset of the harvest season in ponds having heavy cover, water levels 50 cm or so lower than those recommended above can be maintained to furnish more shallow water for young crayfish.

During the harvesting season the water level should be kept fairly stable. This requires frequent pumping to replace water lost by evaporation. Pumping must also be resorted to replace deoxygenated water. Any abrupt reduction in harvest is usually a sign of deoxygenation or some such condition and calls for partial replacement of water.

Good Supply

Any animal food that cultured crayfish get is the result of natural production and their own foraging, but successful culturists encourage suitable food plants and discourage large, tough plants which interfere with harvesting, shade out smaller, more desirable plants, and are inedible by crayfish. Crayfish from ponds that do not contain adequate amounts of edible plants often have brown or black "fat" or livers, and tails which are not filled out, and are considered to be of inferior quality to well nourished crayfish with full tails and yellow livers.

Plants used as food and/ or cover in crayfish ponds must be

capable of survival both while the pond is flooded and when it is dry. The most desired food and cover plant in crayfish ponds is alligator grass *(Alternanthera phylloxeroides)*. It can be seeded into a pond by raking it out of ditches and scattering it in the pond during drawdown in June and July. Alligator grass may, on occasion, grow too thick and hinder harvesting. If harvesting is done by boat, boat trails may be disked or raked on the bottom during the summer. Although cattle should generally be kept out of crayfish culture areas, grazing them on the dry pond bottom may help retard alligator grass.

Water primrose *(Jussiaea* spp.) is now considered as good as or better than alligator grass and may be planted together with it. Water primrose has the advantages of not growing as thick as alligator grass and being more tolerant of cold weather. A number of other plants which occur naturally in crayfish ponds, including pondweeds *(Potamogeton* spp.), *Elodea,* and duckweed *(Lemna)*, are fair food and/or cover plants and should not ordinarily be discouraged. It is certain that fertilization of dry ponds would benefit all of these and other plants, but techniques to encourage desired species have yet to be worked out.

Crayfish will also eat almost any soft terrestrial plant. Some farmers have planted sorghum or millet as feed for crayfish and claim increased yields, but there is as yet no experimental evidence for this conclusion.

It is likely that supplemental feeding with animal matter would increase production, but it is doubtful whether it would be economically feasible. Experiments with feeding young red crayfish at Auburn University showed that growth was better on a diet of *Elodea* and ground *Tilapia* than on either component alone. Research is currently being carried out at Louisiana State University with the aim of developing an economical, high-protein artificial feed for crayfish.

Problems

The most common pest plants in crayfish ponds are cattails *(Typha* spp.) They can usually be controlled only if caught early, at which time individual plants can be pulled out manually. If dense growths occur, disking and raking the pond immediately on drying may be helpful. Otherwise, it is necessary to call on state or federal agencies for assistance. Outside help is also usually necessary to control water hyacinth *(Eichornia crassipes)* which, unchecked, may completely cover the surface of a pond. A number of emergent plants other than cattails are usually present but seldom reach problematical densities in ponds which are drained and dried annually.

Filamentous green algae, which occur principally during the winter, also may hinder harvesting operations. Further, they are prone to die off suddenly in warm weather, creating a pollution problem. Partial removal of floating algal mats may be effected by vigorous circulation of the water .

Man is by no means the only animal which is fond is crayfish, and predation by fish, birds, raccoons, bullfrogs, snakes, turtles, salamanders, large water beetles, and others, is a problem confronting every crayfish culturist. The best protection against most of these predators is abundant cover in the pond, but fish, the most serious predators, may require additional measures. Annual draining and drying of course substantially reduce fish populations, but some of the worst predators, notably green sunfish *(Lepomis cyanellus)*, bowfin *(Amia calva)*, and bullheads *(lctalurus* spp.), may survive in potholes until reflooding time. Rotenone at 2 to 3 ppm will destroy such fish. Great care should be exercised in its use, however, and a biologist consulted if possible, since 5 ppm of rotenone will kill some young crayfish.

It is standard practice to screen pond outlets to prevent fish from entering, but opinions differ as to the feasibility of screening inflow pipes. Many successful operators reason that most large fish are killed by the pump impellers and do not screen inflow pipes. In any event, the large 0.3-to 0.7-m pumps employed on most ponds move great volumes of water and make screening difficult.

It is recommended that, on new ponds supplied by surface waters, a temporary screen be installed on the inflow pipe and checked frequently during the first few days. If considerable numbers of large predatory fish are captured, a permanent screen should be employed. A suitable design consists of a cylinder of 25-cm weldwire about 1 m in diameter and 4 m long. Such a screen will allow young of predatory fish to enter the pond, but finer screens clog up very rapidly and require constant cleaning.

Harvesting

One of the most difficult aspects of pond culture of crayfish is harvesting. The most frequently used harvesting device is an 0.8 to 1.0 m long funnel shaped trap made of 21-cm mesh chicken wire and so constructed that it folds flat for hauling. Shorter traps are fully as effective for capturing crayfish, but dissolved oxygen concentrations may become very low in crowded, submerged traps, causing mortality. Long traps, however, may be propped up at an angle so that a small portion is above the surface of the water, and trapped crayfish can obtain air.

Traps are commonly baited with fish heads or chunks of gizzard shad *(Dorosoma cepedianum)*, which, is considered more effective than any other fish. Fresh fish may be scarce during the winter, and soybean cake, perforated cans of dog food, or almost any high-protein substance may be substituted. About 25 traps/ ha are set out.

Most of the harvesting is done by professional fishermen hired by the culturist. This compounds the culturist's harvesting problems, since he is seldom able to maintain a crew of fishermen throughout a harvest season. But to realize maximum yields and fulfil contracts to buyers it is essential that harvesting be intensive throughout the season. The problem is particularly acute during cold weather, when fishermen may be reluctant to run their traps frequently, and during the height of the wild crayfish season, when they may be able to make more money harvesting wild stocks on their own. The demand for labour is particularly great at the start of the season, in late November and early December, when many culturists prefer to harvest as heavily as possible to beat the peak of the wild crop, to utilize adults which have spawned and will die shortly and reduce the danger of crowding of the young crayfish.

Some farmers circumvent harvesting problems by opening their ponds to the public, either for a set free or by charging substantially less than the retail price for the crayfish which are caught.

Table 9.2 Schedule or Procedures for Pond Culture of Crayfish in Louisiana

Activity	*Dates*
Construction of new ponds and flooding	By May 15
Stocking	June 1-30

Draining	June 15-July 15 (begin 2 weeks after stocking)
Fish control	July 15-August 1
Repairs, improvement, planting, and weed control	July 15-August 15 (as soon as possible after most of the pond is dry)
Reflooding	By September 15
Harvest	November 25-June 30 (begin 1 month later for new ponds)

Rice Field Culture in Louisiana

Crayfish have long been harvested as an incidental crop from rice fields in Louisiana, but it is only in the last 15 years that intensive cultivation has become common. Today, nearly 1000 ha. of rice fields are operated in 2-year rotation as crayfish ponds and pastures. The approximate schedule for this sort of rotation is given in Table 3.

There are usually enough crayfish present in a Louisiana rice field to serve as brood stock. If not, adults may be stocked at 6 to 12 kg/ha during May. Some crayfish leave the drained fields in July and August, but most of them burrow into the moist soil. Rice stubble resprouts about a week after harvest, at which time the fields are reflooded to a depth of 15 to 45 cm. Crayfish in rice fields feed mainly on rice stubble and various aquatic plants which find their way into the fields. Rice-crayfish farmers should not use rice seed treated with aldrin, which may be toxic to crayfish.

Table 9.3 Schedule of Procedures for Rice Field Culture of Crayfish in Louisiana

Activity	*Dates*
Plowing	March 1-April 30
Replowing and planting rice	April 15-May 15
Flooding and stocking if necessary	May 1-May 31
Draining	July 15-August 15
Harvesting rice, followed by reflooding	August 1-September 1
Harvesting crayfish	December 1-June 15
Drain and use as pasture	June 15-March 1

Harvesting crayfish from rice fields, although tedious, is easier and cheaper than in ponds. Due to the more regular bottom, nets may be employed. The usual type, called a drop net or umbrella net, consists of a rectangular piece of netting, each corner of which is connected by a wire or rod to a ring located over the center of the net. A pole or other lifting device is inserted in this ring and the net laid flat on the bottom. Crayfish are baited into it with beef pancreas or other offal and captured by lifting the pole. Ordinary farm laborers rather than professional fishermen may be hired to do this chore.

Growth to the minimum marketable size of 10 to 15 9 is rapid in either type of culture, requiring little more than 6 months. Larger 40-to 45-g crayfish, which bring the best prices, are harvested principally in the early part of the season and may be 8 to 14 months old. Well-managed ponds and rice fields commonly produce 400 to 700 kg/ha of crayfish, and yields of better than 1100 kg/ha have been achieved.

Marketing

Marketing is by no means organized. Crayfish may be sold live, boiled; or as peeled tails to wholesale dealers, restaurants, or individuals. Production or peeled tails may be increased if a mechanical crayfish peeler, currently being developed at Louisiana State University, is perfected.

Diseases and Parasites

Disease is not a major problem to crayrfish culturists in Louisiana. A bacterial rot causing decay of the rostrum was observed in red crayfish in Alabama but was readity cured by treatment with potassium permanganate at 3 ppm. An extremely heavy infestation of both red and white crayfish with larvae of the water boatman *Ramiphocorixa acuminata* occurred at the U.S. Fish Farming Experimental Station, Stuttgart, Arkansas, but this insect has never presented serious problem in commercial culture.

10

Techniques of Oyster Culture

Marine aquaculture may well have begun with oysters, which were cultivated in Europe during Roman times. The origins of oyster culture, as opposed to harvesting "wild" oysters, may be rooted in the human urge for convenience, but there are at least two other historic rationales for the practice:

1. Many areas where adult oysters grow very well are not suited to their reproduction. These areas may be brought into production of edible oysters by simply transplanting excess young oysters from grounds where reproduction occurs.

2. Though oysters may be prepared for the table in a number of ways, many gourmets prefer them raw, on the halfshell. Regularly shaped osyters, with shells free of fouling organisms, are considered more desirable for this purpose. Such oysters are rare in nature but can readily be produced by oyster culturist.

In this century, the trend toward oyster culture has been accelerated as a reaction to man's catastrophic impact on estuarine ecology. In the Orient, human population pressure has stimulated the development of aquacultural schemes for many organisms, including oysters, which are regarded not as a gourmet dish but a staple seafood. This has led to the development of various techniques of off-bottom shellfish growing which have increased production many fold over natural levels.

The ever-increasing food demands of the exploiding human

population might in themselves have resulted in some depletion of natural oyster stocks. But, particularly in the "advanced" nations, population growth has been accompanied by destruction of natural oyster grounds, both intentionally in the course of development of seashore property for other purposes and unintentionally through domestic and industrial pollution. Pollution constitutes a double threat to shellfish industries; even where the animals themselves are not harmed they may, due to their habit of filter feeding, concentrate pollutants in their flesh and become unfit for human consumption.

Pollution effects on shellfish fisheries and culture are most severe in the more developed countries. By 1964 nearly 400,000 ha. of shellfish grounds, or about 12% of the total active producing area in the United States, were closed to harvesting for health reasons. In some states all waters are closed. The net effect of pollution, overharvest, and generally poor management has been to reduce oyster production on the Atlantic coast of the United States from more than 43 million kg in 1920 to less than half that amount in 1964. In New York state, a principal producer or oysters but also a highly populous and industrialized state, oyster production over a 50-year period declined by 99%.

In recent years pollution and other factors have threatened oyster production in many countries. The town of Malebon, once the chief oyster port of the Philippines, has been virtually eliminated from the industry by pollution. Industrial pollution, siltation, and overharvest threaten future oyster harvests in several other areas of the country. In Australia, where oyster culture was initiated in 1896 to compensate for overharvest of wild oysters, the future of the industry in the principal producing area, the George's River estuary, is threatened by domestic pollution.

Some efforts have been made to reduce pollution in oyster-producing areas, but the principal response to the problem has been to seek biological and technological refinement of culture systems, including the construction of hatcheries where all or part of the life cycle can be carried out under controlled conditions, using water of known purity. D.H. Wallace of the National Marine Fisheries Service has suggested that this is the only hope for the oyster industry in the United States.

Natural History of Oysters

Before efforts to conserve or culture oysters or any organism can proceed, the natural history and ecology of that organism must be understood. The American oyster *(Crassostrea virginica)* has probably been studied more thoroughly than any other marine organism, and the reader is referred to P.S. Galtsoff's (1964) monograph for details on the American oyster in nature. Here we briefly describe only those aspects of the oyster's life that are of particular interest to culturists.

Spawning

Spawning commences whenever the water reaches a certain temperature (the precise temperature depends on the species; see Table 10.1) and lasts as long as the temperature remains above that minimum. In the tropics this may result in year-round spawning. In oysters of the genus *Crassostrea,* which includes most of the cultured species, the genital products of both sexes are released into the water and fertilization and hatching take place outside the shell. In oysters of the genus *Ostrea,* the male releases sperm, but the female retains the eggs in her pallial cavity until hatching. Individuals of some species are alternately male and female.

Hatching and Settling

Hatching occurs anywhere from a few hours to more than a week after fertilization. The larvae are free-swimming for a few days to a few weeks, after which they metamorphose, settle on some solid object, and assume a sessile life habit for the rest of their lives. As the newly settled oysters or "spat" develop, mortality from predation, disease, and the effects of crowding may be very high. Development to sexual maturity takes about a year, but up to 4 years are required for wild oysters to attain marketable size. Both larvae and sessile oysters feed by filtering plankton and suspended organic matter.

Environment

Environmental requirements and tolerances for the various species are imperfectly known, as is oyster taxonomy, but Table 10.1 serves as a general guide to the habits and habitats of the most important commercial species. A number of species in addition to those listed in Table 10.1 are locally important, notably the Olympia oyster *(Ostrea lurida)* which brings the highest prices of any oyster in the northwest United States., and *Crassostrea rivularis* of Japan, which, due to its extremely high tolerance for turbidity, is the only oyster that can be cultured over some bottoms.

BASIC TECHNIQUES OF OYSTER CULTURE

History

In nature, oysters settle on any hard substrate. Areas where conditions are well suited to their growth soon become covered with many layers of oyster shells. The most primitive methods of oyster culture involve little more than scattering cleaned oyster shells, called "cultch," on the bottom in such areas just before setting and letting nature do the rest until harvest some years later. Seventeenth-century Japanese culturists refined this method by using rocks, branches, and other objects of a size such that they could easily be moved from place to place with oysters attached.

In 1673 Gorohachi Koroshiya, a Japanese clam culturist, made the serendipitous discovery that oyster spat would settle on upright bamboo stakes anchored in the sea bottom. Not only are oysters so attached protected from benthic predators but the resulting three-dimensional "bed" offers many times more surface area for attachment than a flat bottom. The next logical step was the development of the many types of suspended collectors, hung from floats, which are in use today.

Spot Collection

Whatever method of capture is used, the time of spat collection is crucial to the success of oyster culture. The best time of year for spat collection varies greatly with species (Table 10.1), locality, and annual fluctuations in temperature, salinity, tide, and so on. If collectors are set out too early, large numbers of barnacles and other undesirable organisms with habits similar to those of oysters may be captured. In such major oyster breeding areas as the Gulf of Morbihan in France, Long Island Sound in the United States, and Miyagi Prefecture in northeast Japan, government biologists assist oyster culturists in selecting the best dates for spat collection. By examining the contents of plankton tows, biologists may determine the presence and abundance of larval oysters of various ages and thus forecast peak setting periods a few days in advance. In other regions, culturists must rely on past experience and their own test collections.

Table 10.1 The Principal Cultured Species of Oysters, Their Distribution and Characteristics

Species	*Countries where cultured*	*Spawning season*	*Spawning temperature ("C)*	*Incubation period*	*Duration of larval period and time of setting*	*Depth and tidal zone inhabited*
Crassostrea angulata (Portuguese Oyster)	Portugal, Spain, Atlantic coast of France; experimentally in Japan, Tunisia, and California	Summer	20 or more	–	15-20 days	Intertidal; in estuaries where current is strong
C. *commercialis* (Sydney rock oyster)	Australia, from southern Queensland to eastern Victoria and New Zealand	Summer and fall	Peak at 21-23	6 hours	14-21 days; February-April	From intertidal to 3 m below low tide
C. *emdelie* (slipper oyster)	Philippines	Spring and summer, peak during rainy season (July-August)	30-33	–	7 days	Intertidal
C. *gigas* (Pacific oyster)	Japan, Korea, Taiwan, Pacific coast of United States and Canada; experimentally in Australia, France. Netherlands, Portugal, Thailand, and United Kingdom	Peaks in Japan: May-June in Inland Sea, August-September in North Japan	Begins at 19-20, peaks at 23-25	5-6 hours	10-14 days; peak in August	Intertidal
C. *rhizophorae* (mangrove oyster	Experimentally in Cuba and Venezuela	Continuous peaks May-September in Venezuela	–	–	Continuous setting, peaks in July-August in Venezuela, February-April in Cuba	0.5-3.0 m (intertidal)

(Table 10.1 Contd.)

Species	*Countries where cultured*	*Spawning season*	*Spawning temperature ("C)*	*Incubation period*	*Duration of larval period and time of setting*	*Depth and tidal zone inhabited*
C. *virginica* (American oyster)	Atlantic and Gulf coasts of United States, maritime provinces of Canada; experimentally in Japan and California	Long Island Sound: mid-July-early October; Chesapeake Bay: mid-June – mid-October; South Carolina; May-October; Gulf of Mexico: April-November	Begins at 20	–	10-21 days	Intertidal to more than 30 m; spawning and spat settling most successful in estuaries
Ostrea edulis (flat oyster)	Atlantic coast of France, Spain, Netherlands, Great Britain, Japan, United States (Maine and Pacific coast).	June-September in Morbihan area of France	20 or more	8 days	12-14 days	Little or no intertidal exposure; in estuaries where current is weak

(Table 10.1 Contd.)

Species	Temperature tolerance (°C)	Salinity tolerance (%)	Substrate	Size marketed and growing time to that size
Crassostrea angulata (Portuguese oyster)	About 15–25	Optimum 20-30; fails to reproduce above 34	Various; quite tolerant of turibidity	65 g (including shell); 3 years in France
C. *commercialis* (Sydney rock oyster)	Varies widely	Varies widely	Hard bottom. usually in the shade	85-100 mm (for consumption on halfshell), 2½ years in North, 3½ years in the South; 505 mm long (for shucking), 2-3 years
C. *eradelie* (slipper Oyster)	25-33, possibly wider	Wide, up to 45, spawns at 15	Usually over mud on	75 mm diameter, 6-9 months plants; cultured over sand; very tolerant of silt
C. *gigas* (Pacific oyster)	15-30, optimum for larval development 23-28 best spat sets at 25 or more	Optimum for larval development 23-28 (varies with temperature) best spat sets at 15-18	Any hard substrate	300 g (including shell), 6-12 months (Inland Sea), 18 months (North Japan) half-shell size, 2 years.
C. *rhizophorae* (mangrove oyster)	18.4–34.0	22., – 40 (extremes of short duration); optimum 26-37	Associated with mangrove roots; very tolerant of turbidity	75-100 mm diameter, 5 months
C. *virginiaca* (American oyster)	Larvae develop well at 17.5–32.2	Wide to at least 32; best' larval development above 16.5, some survive at 7.5, maximum for larvae 22.5	Hard	75 mm diameter, 4-5 years, Northern Atlantic Coast; 2-3 years, mid-to South Atlantic Coast and Gulf of Mexico
Ostrea edulis (flat oyster)	Wide-at least maximum 15-20 mortality at variation	Usually found above 25	Hard, not tolerant of turbidity	65 g (including shell), 75 mm diameter, 4 years

Proper placement of collectors is fully as crucial as timing. In many parts of the world, only a small percentage of the oyster grounds are suitable for spat collection. Thus a few fortuitously located culturists may specialize in the production and sale to other culturists of young oysters. Or oyster growers may collect spat on grounds far removed from their growing areas.

Yet another important factor in spat collection is the type of collector used. Dozens of materials and designs have been tested, but no one type has proven universally superior. The best collector for one species in one area may be a failure with another species somewhere else. Therefore the various collectors are discussed here under the descriptions of individual culture methods.

Spat are usually left on the collectors at least until they develop into "seed" oysters, 10 to 20 mm in diameter. Seed oysters may be cultured in place, moved to areas best suited to growing for market, or sold to other culturists.

Growing for Market

Procedures used in growing adult oysters for market vary with species and locality and according to the use for which the oysters are destined. Oysters to be eaten raw on the halfshell must be given special care to insure the production of uniformly shaped shells if they are to bring the best prices.

The same principle that favours setting of spat on three-dimensional hanging collectors favours off-bottom culture of oysters for market. Nevertheless, bottom culture, with its inefficient use of the water column and the attendant dangers of siltation and predation, is still practiced in some countries, notably the United States, Canada, and France. In the United States and Canada it persists due to legal obstacles to leasing portions of the sea bed, as is done in other countries. In North America, most coastal waters are held to be public domain, and to place extensive floating structures in them would interfere with recreation as well as or commercial use, whereas bottom culture does not raise such problems.

OYSTER CULTURE ON THE ATLANTIC COAST OF NORTH AMERICA

Spat Collection

Though market size oysters are rarely grown off bottom in North

America, spat collection, which takes up considerably less space than growing for market, may be carried out using collectors suspended from rafts. This method is particularly applicable where oyster spawners exist in brackish ponds or lagoons, which may not be considered public waters. The most notable example of this practice is the Ocean Pond Corporation, located on Fisher's Island in the New York waters of Long Island Sound, a body of water where oyster spawning is undependable at best. Ocean Pond is a 9.3-ha body of brackish water connected to the sound by a narrow channel. Annually, just prior to the oyster spawning season, this channel is blocked to prevent escape of oyster larvae and to accelerate warming of the water. The abundance of larvae is determined by continuous monitoring and when adequate quantities are detected collectors composed of scallop *(Pecten)* shells strung on 2.4-m long galvanized wires are suspended from rafts. When setting is completed the pond is opened to the sea, but the juvenile oysters are not moved until the following spring, when they are sold to oyster growers throughout Long Island and New England. As of 1969, Ocean Pond Corporation had experienced successful sets for seven consecutive years, sustaining an annual production of 25 to 50-million 9-month-old seed oysters.

The success achieved by Ocean Pond Corporation is by no means typical in the northeast United States and eastern Canada, and inadequate supplies of seed oysters remain a problem for the industry in that region. Attempts have been made to import seed from the southeastern states, where spat setting is more reliable, but southern strains of C. *virginica* do not do well in the cold waters of the northeast.

Biologists in Maine have had sporadic success in spat collection and efforts to take advantage of natural spatfalls in northern waters continue in several areas. The most successful collectors in most cases have been strings of scallop shells. Cement-coated egg cartons proved a satisfactory substitute for shells in Prince Edward Island. In a few localities, chicken wire bags containing shells have proven more efficient than strings, but the opposite is generally true. Various kinds of artificial collectors have also been rested but have generally proven inferior. However 3 mm thick sheets of such nontoxic plastics as polyethylene or polypropylene have shown promise. Collectors made of these plastics are much lighter and easier to handle than strings or bags of shells and are cheaper to construct than shell strings. In recent years culturists in California, Virginia, and Canada have had good results using plastic mesh collectors similar to the "Netron" collectors used by Japanese culturists (see below). These have a special advantage for bottom culture in that the seed oysters can easily be removed.

Since most oysters grown north of New York are declined for consumption on the halfshell, crowding occasioned by bunching on the collectors is to be avoided. For this reason some culturists prefer bags of shells to strings since if an excess of spat is collected, it is a simple matter to break shells or separate them one from another. Synthetic collectors may be manufactured with the same requirements in mind by simply alternating bands of smooth material, from which spat may easily be displaced, with bands of roughened material.

Hatcheries

Discussion of spat collection is academic, however, if adequate numbers of larvae are not present in nature. Since shortages of natural spawn are the rule in the northeast, the emphasis among successful culturists is on hatchery production of seed oysters.

Hatchery culture of oysters had its genesis in 1879 when W.K. Brooks of the Johns Hopkins University demonstrated that American oyster eggs could be hatched in the laboratory. Success in rearing larvae to the setting stage was not achieved until 1920 when W.F. Wells of the New York Conservation Commission opened the door for practical intensive oyster culture. In the mid-1940s a modification of Wells' technique, for the first time incorporating artificially induced spawning, was developed by V.L. Loosanoff and H.C. Davis at the U.S. Bureau of Commercial Fisheries Biological Laboratory at Milford, Connecticut. One or both of these techniques, known respectively as the Wells-Glancy method and the Milford method, are used at the five commercial oyster hatcheries currently operating on Long Island Sound.

One of the most successful hatcheries, located at the time at Oyster Bay, New York, produced 100 million seed oysters in 1966, using essentially the Wells-Glancy method. That hatchery has since been relocated and expanded and its methods somewhat modified, but the essential elements of the practice remain the same and are described here as carried out in 1968. Sexually mature adult oyster are selected from the growing beds (the selection based on size, shape, and growth rate) and are held in the hatchery at 10°C They are then conditioned for spawning by slowly raising the temperature to 18°C or more and holding the oyster at that temperature for 2 to 4 weeks, the duration depending upon the time of year. Spawning is then induced by raising the temperature to 25°C while holding the oysters in glass trays. By using this technique it is possible not only to spawn oysters on any day of the year but, more important, to spawn individual oysters twice a year or perhaps more frequently.

The fertilized eggs are transferred to one of a battery of 50 120-gal conical rearing tanks. Every other day the water is drained through the bottom of the tank, the newly hatched larvae being caught in a fine mesh screen. After each change of water the tanks are scrubbed with Alconox or Clorox to guard against disease organisms. The larvae are then graded by further screening, keeping only those individuals over 0.3 mm in diameter, or about 20% of the total, and discarding the remaining 80%. This process is believed to be very important in that it selects for a strain which retains the characteristic of rapid growth throughout life.

The selected larvae are then transferred to another tank containing water pumped into the hatchery from Oyster Bay, centrifuged through a battery of three Sharples centrifuges, and pumped into four 20,000-liter tanks in the upper floor of the building. A greenhouse-type roof admits sunlight and keeps the temperature above 30°C The centrifugation of the water removes the animals and larger algal cells. If a sufficient number of small, unicellular algae suitable as food for the larvae are present and escape centrifugation, these are allowed to grow naturally for 24 hours; if such microoganisms are scarce, the tanks may be inoculated with 200 litres of algal culture and fertilized. The resulting culture of microorganisms, natural or inoculated, is then used to fill the larval rearing tanks.

After 10 to 15 days the larvae are ready to set, and when this is confirmed by microscopic examination they are transferred to one of six 3600-liter plastic settling tanks, each of which contains 10 bushels of specially selected, screened, and washed oyster shells, which may be treated with Clorox as an added precaution. It is most important to use clean shells in this operation or a bacterial bloom may develop in the tanks and kill the larvae or prevent setting. The clean shells are spread in the bottom of the tanks and the larvae introduced and left for 24 to 48 hours, during which time setting occurs.

Spat so collected are not as convenient to handle as spat set on hanging collectors. A proposed method of collecting spat in hatcheries involves releasing ready-to-set larvae into a long, narrow tank. After releasing the larvae, special collectors made of nontoxic plastic and almost the width of the tank would be passed through the entire length of the tank at a speed regulated by the rate of spatfall. With supervision by an experienced person, this method might prove highly efficient.

The shells with the spat attached are then transferred to half-bushel plastic mesh bags which are suspended from wooden beams into the nursing tanks. The 27,000-liter cement nursing tanks, each of which holds 200 bags, are located in a greenhouse where the temperature iş kept at or near 30°C. These also are fed with water from one of the roof tanks in which an algal bloom is maintained, one-half of this tank each day being added to each holding tank. The high temperature and supplemental feeding results in rapid growth of the young seed oysters. The length of time they are kept in the holding tanks ranges from 4 days to a week or more, depending upon the availability of space and the timing of subsequent operations.

During this stage fouling by such animals as mussles, annelids, tunicates, gastropods, and hydroids may occur. At the Milford laboratory this was prevented by periodic treatment with Victoria Blue B, a biological stain, Clams, which customarily keep their shells closed during treatment, are unaffected by long-term exposure to low concentrations of this stain. Oysters, however, may open their shells and suffer mortality. Thus they are best treated by dipping them in a high concentration, say 200 ppm, for 15 min, which should kill most fouling organisms.

Eventually, the wooden beams with shell bags attached are transferred by chain-hoist and overhead rail outside to the dock where they are placed on floating rafts, each of which holds 200 bags. The seed oysters are suspended from the rafts into the bay water for 2 to 3 additional weeks until they reach fingernail size (1 to 2 cm), when they are ready for planting on the oyster beds. At full operation, 16 rafts or 1600 bushels (1 bushel=35.2 litres) of seed bearing shell are in use in the bay at one time.

Not all culturists introduce young oysters to bay water so soon after setting. Some growers are attempting to keep spat in large tanks for longer periods in order to enhance growth and reduce predation and fouling. Protection may also be afforded by placing the spat in natural or artificial ponds it should be pointed out that artificial ponds may be relatively sterile, resulting in poor growth), intertidal waters, or the upper reaches of estuaries, where salinity is low, thus predation and competition are minimal. In such environments, mild sewage pollution may have the beneficial effect of producing food in the form of rich plankton blooms.

The entire operation from spawning to fingernail size seed requires 4 to 6 weeks. The hatchery can produce the small (3 to 4 mm) seed throughout the year, but the additional growth to 1 to 2 cm on the rafts cannot be done after the water temperature falls below about 10°C so the seed oysters are not planted on the beds in late fall or winter when they cannot grow and are at their most critical period for survival. Thus the hatchery operation has been restricted to late spring and summer.

The Milford method is similar to the Wells-Glancy method in most respects but relies on pure algal cultures as a food source. Maintaining algal cultures is somewhat more expensive than centrifuging sea water but has the advantage of allowing the culturist to screen out such undesirable forms as *Chlorella,* which produces toxic metabolites, in high concentration fatal to oyster larvae, and to select for forms readily digested by the larvae. Among the latter are such chrysomonads as *Monochrysis lutheri* and *Isochrysis galbana,* which do not have thick cell walls. Another suitable species is *Scenedesmus obliquus* originally imported from Japan.

The advantages of pure algal cultures are not so clear-cut as they might seem, however, since we know little of the food habits of oysters in nature and even the best pure cultures are likely to be lacking important components of natural diets. We do know that no one species of algae in itself constitutes a satisfactory diet for oyster larvae.

Another point in favour of pure algal cultures as a food source is that the population density of larvae in culture may be maintained at a higher level. No matter what steps are taken to ensure an algal bloom, the amount of usable algae in a given volume of sea water is limited. Thus as the larvae grow they must be periodically thinned.

If a pure algal culture is used as food, however, dense populations of larvae may be maintained simply by increasing the rate of feeding.

Some Wells-Glancy type hatcheries make occasional use of pure algal cultures when natural populations of algae are deficient and/ or when light and temperature conditions are not adequate for the development of sufficiently dense cultures in greenhouse tanks. At one successful hatchery, algal culture is carried out in a constant temperature (about 20°C) room illuminated with fluorescent lamps. Sea water pumped into the room is filtered through wound-orlon filters and then passed through a bank of 4 2-ft ultraviolet lamps to kill any remaining microorganisms. The water is then enriched and used to fill the culture vessels, which consist of 11 227-liter and 8 136-liter translucent plastic drums. Presently in use as food organisms are the flagellates *Isochrysis galbana, Monochrysis lutheri,* and *Rhodomonas* sp.

As needed, these cultures are fed to the larvae to produce a total ration of 2000 cells/(larva) (day). The same concentration is used throughout the larval cycle but larger algae are used as the larvae grow larger .

The rationale for the construction of oyster hatcheries in North America is that in many years natural reproduction in certain areas does not provide enough larvae to sustain commercial culture, but development of hatchery techniques also opened the door for selective breeding of the American oyster.

The critical factor in determining the success or failure of an oyster hatchery is water quality. To insure good survival of larvae the water should have a salinity of 16.5% or greater, a pH of 6.75 to 8.75, and be relatively free of pollutants, particularly metallic salts, pesticides, and detergents. Areas where intensive, possibly toxic algae blooms occur frequently should be avoided. Silt may also have an adverse effect on oyster larvae, directly and by alteration of pH, but can be relatively easily removed by filtration. Even if the water quality is excellent, it may be necessary to treat incoming water with antibiotics, sulfa drugs, or ultraviolet radiation. To produce 5000 bushels of spat requires 32,000 to 40,000 litres of cleaned sea water daily.

A recent innovation which is finding increasing use by oyster culturists is the hatchery production of "cultchless" or "free" spat by inducing the oysters to set on sheets or screens of plastic or metal from which they may be removed, or on tiny particles of calcium carbonate. The free seed may then be grown in fine-mesh screen trays, in the hatchery or out-of-doors, until they are large enough to be planted on the bottom or, in some cases, until they reach marketable size. The advantages of this technique is that many more free seed oysters can be reared in the same space, they can be handled and transported far more easily and less expensively, and they grow into much more evenly and regularly shaped adult oysters than oysters which have set on oyster or scallop-shell clutch.

Growing for Market-Bottom Culture

Although hatchery production of seed oysters on Long Island Sound is highly sophisticated biologically and technologically, techniques used in growing seed oysters to marketable size in the United States barely qualify as culture. As with wild oysters, growth and survival of cultured oysters is largely determined by the quality of the beds, which should be located on a hard bottom in 1 to 12 m of water where tidal currents are strong. Once seed oysters are planted on the beds management is confined to predator and siltation control, and when necessary, thinning and movement to new beds. One successful grower routinely thins oysters twice during the growing period. Seed oysters planted on the beds at about 200 bushels/ha (1 bushel=35.2 litres) are redistributed over five times as much bottom at the end of the first year of growth, by which time they amount to 2000 bushels. At the end of the third year they have reached 4000 bushels and are distributed over twice as much bottom as at the end of the first year.

Even more primitive methods are used maintaining public oyster grounds in the eastern United States. The various state governments initially place shells on the bottom in areas where reproduction occurs, then move them, with spat attached, to more saline waters for growing. Up to three subsequent moves may be undertaken, but little else is done by way of management except to regulate the season and method of harvest.

Accumulation of silt is particularly dangerous to young oysters, which may be smothered or prevented from feeding. Proper selection of beds plays a major role in discouraging siltation, but frequent transplantation and/ or supplemental control measures are usually practiced by commercial culturists. The larger growers customarily clean oyster beds of silt using suction dredges or high-pressure water jets.

Predation is a problem wherever oysters are grown on the bottom, but it has been particularly severe in Long Island Sound, where the chief predators are oyster drills *(Urosalpinx cinerea* and *Eupleura caudata)* and the common starfish *(Asterias forbesi)*. The upward trend in oyster production in that body of water since 1965, after 50 years of decline, is considered to be largely due to effective control of these animals. Long Island culturists, who formerly expected 10,000 bushels of seed to yield 10,000 bushels of oysters, now produce 100,000 to 150,000 bushels from the same amount of seed. Unfortunately, one of the basic techniques of predator control relies on chlorinated hydrocarbon biocides. These compounds, already beginning to be seen as ecological hazards on land, are even more frightening when applied to culture of aquatic filter feeders, which are extremely efficient in concentrating waterborne chemicals. But thus far, despite the history of DDT and other "safe" pesticides, the philosophy of the shellfish culturists seems to be "If it hasn't been *proven* dangerous to man, it's all right."

The principal biocide used in oyster culture in Long Island Sound is a potent mixture of chlorinated hydrocarbons marketed as "Polystream," which is illegal in most states. A week, before stocking young oysters, drills (as well as starfish and other predators and competitors) are removed from the beds by suction dredging. Then Polystream is applied at 2000 kg/ha. Such treatment effectively curtails predator populations for 1 to 2 years and has been largely responsible for reducing annual losses to drills from 50% or more to less than 1 %.

There are other, more laborious methods of effectively controlling oyster drills. One of the simplest involves use of an underwater plow, adapted from an agricultural disk plow. If the bottom is suitably soft, this device may be used to bury and kill drills, starfish, and perhaps limpets and mussles. In experiments conducted by the U.S. Bureau of Commercial Fisheries, drills buried under 6 cm of sand or mud suffered 92% mortality in 5 to 55 days. The higher the temperature, the more quickly they were killed.

Various chemicals less hazardous than the chlorinated hydrocarbons may be used as dips to kill drills and other predators and competitors, but their use entails the expense and labour of dredging up the oysters. Of these substances, rock salt is easily the cheapest as well as one of the most effective. By dipping the oysters in a saturated solution, then exposing them to the air for a while, most animals other than shellfish are killed, but oysters are unharmed. The time of immersion in the salt solution and exposure to air varies according to the type of organism to be controlled (Table 10.2).

Table 10.2 Immersion Time in Saturated NaCl Solution Required to Kill 5 Types of Oyster Predators or Competitors

Predator or Competitor	*Immersion Time (Saturated NaCl) (Min)*	*Air Drying Time*
Oyster drills	5	Several hours
Starfish	1	Several minutes
Boring sponges (*Cliona* spp.)	3	1 hour
Limpets (*Crephidula* spp.)	3-5	30 min-1 hour
Tunicates	1-3	1 hour

The salt dip technique, with suitable modifications, is also likely to prove effective for controlling the predatory flatworm *Stylochus ellipticus* as well as many fouling organisms, including protozoa, hydroids, bryozoa, crustaceans, and algae. Small mussels are usually unaffected by salt treatment but may be controlled by similar treatment with a 0.5 to 1.0 solution of copper sulfate. A quick dip, followed by air drying for 24 hours, is effective against not only mussels but a number of other undesirable organisms, notably drills. If the strength of the solution is increased to 2.0%, the drying time in air can be cut to a few hours.

Drills are among the deadliest enemies of untreated oysters, but starfish, which have been estimated to destroy more than 500,000 bushels of oysters annually in Long Island Sound alone, require more vigilance on the part of the culturist. Even when eliminated from an oyster bed, they may rapidly reinvade. To guard against starfish and other hazards, including siltation, scuba divers may be employed to continually monitor oyster beds. This precaution has played an important role in increasing the survival rate from seed to market size of one firm's oysters from 1 to 70%.

Constant treatment with suction dredges also plays a role in the fight against starfish. An older and more widely used device for the same purpose is the starfish mop, which consists of an iron beam 2 to 4 m long with a number of light chains attached. At the end of each chain is a bundle of rope yarn about 2 m long. As these mops are dragged slowly over the oyster beds the starfish become entangled in the yarn and are taken aboard and killed by immersion in boiling water.

Another popular control method involves quicklime which, when it comes in contact with a starfish, causes rapid disintegration and eventual death through its caustic action on the body membranes. Lime is usually applied at about 300 kg/ha, but up to 500 kg/ha may be used on beds located in deep water. To achieve complete coverage and to insure that the lime will adhere to the starfish very fine particles are used. This necessitates pumping it to the bottom so that slaking will not occur before the lime reaches the oyster bed. Though very effective, this practice is relatively expensive and may prove harmful to other commercially valuable animals such as lobsters.

Efforts have been made, without success, to institute biological control of starfish. The possibility of harvesting them for use in the manufacture of protein concentrates has also been explored. Starfish protein concentrate compares favourably with good quality fish meal, but the rate of natural production of starfish is not sufficient to support a profitable fishery. Until a cheap and efficient biological or mechanical method of starfish and drill control is developed, chemicals, however hazardous, are likely to continue to play a major role in bottom culture of oysters in North America.

The techniques used in oyster in the Gulf of Mexico and elsewhere along the Atlantic Coast are essentially the same as those used in Long Island Sound. From New Jersey south, however, osyters are customarily harvested after only 3 years growth. Formerly they were grown for the same length of time as northern oysters, but in the last 10 to 15 years a number of diseases have appeared which, until cures or resistant strains of oysters are found, necessitate early harvest to avert heavy losses.

In some places in the south oysters have been successfully grown together with shrimp and various species of fish in fertilized brackish water ponds. It is not known whether the oysters derived any benefit from fertilization, but experiments in organic fertilization of ponds used primarily for oyster culture indicate that the American oyster derives little food from the type of algal bloom created by adding common fertilizers to sea water.

Crude as American oyster culture methods are, the yields achieved by culturists are many times higher than those recorded for public oyster grounds, which receive negligible management from government agencies. Prior to 1966, the better growers in Long Island Sound obtained yields of about 1000 kg/ha/year, whereas the average yield of public grounds was 10-100 kg/ha/year. Adoption since 1966 of improved methods of controlling siltation and predation has enabled Long Island Sound culturists to achieve a significant increase in production, so projected yields in that area now are of the order of 5000 kg/ha/year.

Growing for Market-Experimental Off-Bottom Culture

High production potential is the principal factor stimulating American oystermen to pursue the possibilities of off-bottom culture, despite the legal difficulties. Seed oysters stored on the bottom over winter were carried through their first summer in floating trays then grown to market size on the bottom. The method was judged biologically effective but economically unfeasible. A similar technique, using strings of oysters suspended from rafts instead of trays, was tested by the U.S. Bureau of Commercial Fisheries in 1956 in Oyster Pond River, Chatham, Massachusetts. They succeeded in producing 70% marketable oysters in 2½ years as compared to the 4 to 5 years required by bottom culture. Further experiments by local shellfish officers in the same area had the extraordinary result of producing more uniformly shaped oysters on suspended strings that were grown on the bottom.

The present center for experimental off-bottom oyster culture in North America is the U.S. National Marine Fisheries Service at Oxford, Maryland. There various types of rafts, long lies, trays and other suspension methods are being tested in the hope of finding biologically, economically, and legally feasible techniques for off-bottom culture of the American oyster in North American waters.

OYSTER CULTURE IN JAPAN

Environmental Requirements

In contrast to American practices, perhaps the most

sophisticated and productive oyster culture in the world is that practiced in Japan. At least eight species are involved, but by far the most important is the Pacific oyster *(Crassostrea gigas)*, which is also an important species for culture. Waters to be used for culture of this species must meet the following requirements:

1. The farming area must be naturally sheltered against violent winds and waves or easily protected by inexpensive construction.
2. Tides and/or currents must be sufficient to change the water completely and frequently.
3. Salinity and temperature must be suitable; optimal conditions are 23 to 28% and 15 to 30°C.
4. The water must contain adequate amounts of phytoplankton as feed for the oysters.
5. The culture area must be free of industrial and domestic pollutants.

Growing for Market-Raft Culture

Some Japanese culturists collect their own seed, but most purchase seed from seed oyster specialists located in the northern part of the country. Various types of bottom culture were formerly employed, but now all oysters grown in Japan are cultured off-bottom, usually suspended from rafts.

In raft culture, 1 month-old seed oysters, about 12 mm in diameter, attached to oyster or scallop shells strung 20 cm apart on No. 13 galvanized wires, with bamboo or plastic spacers, are suspended from the rafts for growth. The length of the strings of oysters, or "rens," depends on water depth in the culture area; in the Inland Sea, this varies from 10 to 15 m.

Rafts used in the Inland Sea are standard in size and construction, being made from 75 to 100 cm diameter bamboo (occasionally cedar) poles which are lashed together with wire in two layers at right angles to each other and with the poles 0.3 to 0.7 m apart. The standard raft is about 16 x 25 m in size and carries 500 to 600 wire rens.

The rafts are buoyed by hollow concrete drums, tarred wooden barrels, or specially constructed styrofoam cylinders. Styrofoam cylinders are replacing the other types of float and are now used in all newly constructed rafts. As the growth of oysters proceeds and the weight increases, additional floats are added as required. In some cases the styrofoam floats are encased in a large polyethylene bag to protect them from barnacles and other fouling organisms. The rafts float at sea surface with no attempt made to keep them out of contact with the water. As a general observation, the Japanese rafts are relatively crude and do not compare with those used in Spain for mussel culture with respect to size durability, or care of construction.

Rafts are commonly laid out in lines 1.6 to 3.0 m apart, tied together with ropes, with two anchors at each end of the line. Ten or more rafts are to become an important crop in the tropics, methods will have to be found to culture native oysters rather than relying on introduction of exotic oysters as was done to initiate the important oyster industries of southern France and western North America. In one instance where introductions were attempted, flat oysters stocked in Bizerta Lake. Tunisia, experienced 100% mortality when the water temperature reached 26°C Portuguese oysters were able to survive but could not reproduce in the highly saline water.

Nigeria

Further south on the African continent experimental oyster culture in Nigeria is resuming after the civil war. Species under study are *Ostrea gasar* and *Ostrea tulipa.* Spat fall of one or the other species has been observed in 10 of the 12 months.

The Caribbean

Continuous spatfall is characteristic of *Crassostrea rhizophorae* of the Caribbean, at least in the southern portion of its range. Wooden collectors appear to be best for this species; investigators in Venezuela and Cuba tested several different types and found respectively that wooden planks painted with bitumen, and branches of the red mangrove *(Rhizophora mangle),* on which C. *rhizophorae* is most often found in nature, were most effective.

Biologists of the Marine Research Station of the LaSalle Foundation at Isla Margarita, Punta de Piedras, Venezuela, estimate that two harvests per year of C. *rhizophorae* are feasible there. Slightly earlier work by the Cuban National Institute of Fisheries indicated a potential production of 6,600 kg meat/(ha)(year) by hanging culture of this species.

Attempts in Cuba to produce a high-quality halfshell oyster by using French methods failed. In addition to problems created by extreme variations *in* salinity (22 to 40%) and an abundance of fouling organisms, C. *rhizophorae* confined in shallow parcs produced such amounts of nitrogenous wastes as to severely pollute their environment.

PROSPECTUS OF OYSTER CULTURE

Bottom Versus Off-Bottom Culture

As Table 10.3 indicates, hanging culture, either from rafts or long lines, is essential if the maximum potential of oysters as a source of food is to be realized. The infinitesimal yields from unmanaged public grounds and the United States, where oysters once constituted the most commercially valuable fishery, suggest that the contributions to fisheries of oysters in highly developed countries are likely to be negligible; oysters must be cultivated if they are to make a significant contribution to human nourishment.

Expansion into New Areas

Future growth of the oyster industry will involve both improvements in efficiency of existing culture systems and expansion into new area of the world. Expansion will entail both development of culture methods for new species, particularly in the tropics, and wider introduction in temperate waters of species for which culture methods have been worked out, notably the Pacific oyster and the flat oyster, which not only possess considerable advantages for culture but are backed by strong seed oyster industries in Japan and France, respectively.

Improvements may continue to be made in bottom culture for some time, and a few new areas may be brought into production using rather primitive methods. For example, in Florida, where American oysters, locally known as coon oysters, commonly form sizable reefs, both construction of artificial reefs made of oyster shells and cutting gaps in existing reefs which have grown above the water surface show promise of increasing oyster fishery yields. However, even in the United States and Canada, where bottom culture is traditional, most current research in oyster culture deals with off-bottom culture. With the exception of such specialized practices as the claire method of producing gourmet halfshell oysters in France, bottom culture appears to be on the way out.

Table 10.3 Average Tield of Different Methods of Oyster Culture in 5 Countries (Meat Weight, Shells Excluded)

Country	*Species*	*Growing Method*	*Production (kg/ha/Year)*
Australia	*Crassostrea commercialis*	Rack culture (best areas)	2,000
	Crassostrea commercialis	Tray culture (best areas)	5,400
Cuba	*Crassostrea rhizophorae*	Experimental raft culture	6,600
France	*Ostrea edulis*	Bottom culture in parcs	250
	Crassostrea angulata	Bottom culture in parcs	1,000
Japan	*Crassostrea gigas*	Long-line culture	26,000
Japan (north)	*Crassostrea gigas*	Raft culture	1,000
Japan (south)	*Crassostrea gigas*	Raft culture	20,000
United States (Atlantic coast)	*Crassotrea virginica*	Bottom culture, with intensive management	5,000*
	Crassotrea virginica	Public grounds-little or no management	10-100

* Projected yield using new methods.

Mechanization of Processing

Improvements in processing as well as culture may contribute to the growth of the oyster industry. Notable in this respect is the development in the United States of a mechanical oyster shucker capable of processing 60 oysters/min as opposed to an average rate of 8/min by hand shucking.

Growing in Heated Water

Use of thermal effluent in oyster culture presents possibilities for both hatchery use and market production. Both possibilities are being tested at Northport, New York, on Long Island Sound, where a major grower has obtained use of a lagoon into which large volumes of heated water are discharged by a Long Island Lighting Company power plant. Since the American oyster feeds and grows only at temperatures above 10°C, the effect of the plant is to provide water suitable for growth on a year-round basis.

Experimental culture of young oysters in the lagoon produced more rapid growth than normal at all temperatures up to and including the mid-summer maximum of 32°C, thus allaying any fears that the thermal effluent might be damaging to the oysters. Since the cooling water is taken from an area of high nutrient content, the high temperatures not only speed up the oysters' metabolism but provide an abundant food supply by inducing a luxuriant growth of microsocopic algae. The shape and location of the lagoon are such that thermal effluent is distributed evenly, thus many rafts or trays, capable of supporting millions of seed oysters, may be set out. To further take advantage of the situation, a hatchery has been built beside the lagoon.

The size of the lagoon is probably prohibitive for large-scale production of marketable oysters but the company, looking ahead to the time when larger power plants will release greater volumes of thermal effluent in less restricted waters, is experimenting with growing oysters to market size in the lagoon. It is believed likely that growing time can be reduced by a year or more.

Selective Breeding

Perhaps the most exciting recent development in oyster culture is the increased effort devoted to selective breeding, which is likely to result in oysters becoming the first truly domesticated aquatic invertebrates. This work is barely begun, but for the most part results are encouraging-rearly unsuccessful attempts to breed strains of *Crassostrea virginica* resistant to the fungus *Dermocystidium marinum* notwithstanding. Among the qualities sought are the ability to reproduce at low temperatures, disease resistance, more rapid growth, and better shape, texture, and flavour.

The Oyster Research Institute at Kesennuma, Japan

Selective breeding has only become possible since the perfection by Loosanoff and Davis of oyster hatchery techniques. Another pioneer hatcheryman was Takeo Imai, former director of the Oyster Research Institute on Mohne Inlet at Kesennuma, Japan. In addition to native oysters, Imai experimented with the American oyster, the Olympia oyster, the Portuguese oyster, and the flat oyster. There are minor variations in culture of the different species, but essentially the same techniques are used for all. The implication for commercial oyster culture is that by using the standard methods developed by Imai almost unlimited numbers of juveniles of any species of oyster may be provided through hatchery rearing.

The facilities of the Institute consist of a small laboratory building, a workshop, and field of 180-1000-liter tanks, which are located in Mohne Inlet. The tanks, which are the heart of the operation, consist of 0.1 mm-thick polyethylene bags 2 m x 1 m wide, about 75 cm deep and are set in a large floating platform approximately 30 m from shore. Seawater is pumped from the Inlet to the shore installation and passed through a deep sand filter. It is then run by hose out to the tank area and passed through a cartridge filter of diatomaceous earth at the hose nozzle before being used to fill the tanks.

The tanks are used for the spawning of adults and the rearing of larvae to the settling stage. Since the tanks are immersed in the sea, there is no temperature control and the organisms grow under the natural temperature conditions of the bay. Thus cultivation is a seasonal operation, with spawning and larval rearing confined to late spring, summer, and early fall. The water temperature of Kesennuma Bay ranges from 5 to 6°C in winter to 24 to 26°C in summer, the latter at the surface only. Much colder water is available in summer in the deeper waters of the bay, including the tank area itself which, though close to shore, is about 10 m deep. Thus a considerable range of temperature is available for different purposes in summer. Salinity of the bay water averages about 33%. This may drop to half this value or less at the very surface during periods of heavy rainfall, but the deeper water maintains its high salinity, so the surface dilution does not present a problem. Water used for filling the tanks is taken from beneath the surface and is therefore relatively constant in temperature and salinity during the summer months.

Some degree of control in the spawning periods of the shellfish can be achieved by keeping the sexually mature individuals in small tanks with refrigerated water. This effectively postpones or prolongs the breeding period of those species which normally commence reproduction early in the season. Spawning in many of the species may be induced by temperature shock, that is, subjecting them to a sudden increase in temperatures of about 5°C for several hours. Frequently, spawning will proceed after the animals so treated are returned to their normal temperature. This treatment is very simply achieved by maintaining the animals to be conditioned in one of the tanks with slowly running sea water, from the supply described, which is first passed through a length of plastic garden hose coiled on top of the tank in an area exposed to the sun. Solar warming of the circulated sea water is adequate to provide the increased temperature for induced spawning.

The adults and larvae reared in the tanks obtain no food from the filtered sea water and must therefore be fed artificially. In the laboratory building there are two rooms devoted to the culture of unicellular algae in 100 or more approximately 14-liter pyrex carboys. The cultures are illuminated with banks of fluorescent lamps and are aerated. To facilitate their preparation, a large volume of water is enriched with culture medium and then passed slowly through a tube exposed to ultraviolet radiation and into cleaned carboys. This does not ensure sterile conditions, but it is sufficient to kill the algae and animals in the filtered sea water and thus insure unialgal conditions. Organisms now in use include the flagellates *Isochrysis galbana* and *Monochrysis lutheri*, and the diatom *Chaetoceras calcitrans*. Stock cultures of other species are available and may be used from time to time. Usually, the larvae are fed the flagellates in the early stages and the diatoms of a mixture of the two in the later and larger stages. Feeding is calculated at a rate of 5000 to 15,000 algal cells per larva twice a day.

Support for the Institute originally came from local and federal fishery agencies and the fishery industry, but as operations become more efficient it is possible to finance more of its activities by sale of juvenile shellfish to commercial culturists. Eventually it is hoped that the Institute can become self-supporting.

The Institute is also doing some research on growing oysters, particularly flat oysters to marketable size. By suspending flat oysters from rafts in 0.3-m x 1-m rectangular metal frames a marketable halfshell oyster can be produced in 2 years, or half the time required by commercial growers in France. At present, however, this method is prohibitively expensive for practical application in Japan.

It would be misleading to imply that the success achieved by the Oyster Research Institute in the rearing of young shellfish could quickly or easily be duplicated elsewhere, for its accomplishments reflect the unique knowledge, experience, and skill of Imai. Nevertheless, the techniques involved are not held in secrecy and there is no reason they cannot, with the help of adequately trained individuals, be adapted and applied elsewhere, as they must be if oyster culture is to progress rapidly.

Increasing Pollution Problems

With continued improvement in hatchery techniques and off-bottom culture methods and the emergence of selective breeding, the short range outlook for oyster culture is one of rapid growth. However, while fisheries and aquaculture in general are threatened by increasing pollution of aquatic environments, oyster culture and other shellfish industries are in a particularly precarious position. Not only are shellfish an estuarine crop, thus more often subject to great amounts of pollution than freshwater or pelagic animals, but as mentioned previously, they tend to concentrate pollutants. Thus levels of pollution far below those required to produce mortality or physiological damage may eliminate oyster culture by rendering its product unfit for human consumption. If pollution were to thus completely remove from production such major oyster growing areas as Long Island Sound, Manila Bay, or the Inland Sea of Japan, it would more than offset all the advances in oyster culture made since the Romans. Even thermal pollution, presently seen as a potential boon to oyster culture, might contribute to such a pollution disaster through synergistic effects with domestic and, or industrial pollutants.

11

Culture of the Indian Carps

It has become a cliche that the people of India need protein. Even the most socially unaware persons have some grasp of the magnitude of India's problem. One can cite statistics *ad infinitum* to demonstrate that India is in a steadily worsening state of nutritional crisis, but however one defines the problem, it is clear that the solution, if there is ever to be one, will not come as any single panacea, but from effective population control coupled with increased and more efficient protein production by many means, including fish culture.

Fish culture shows promise in India, for there is a considerable demand for fish (total landings of fish in 2003 amounted to 7331 thousand metric tons), a number of native fish well suited to culture; a tradition of fish culture, and a great abundance of cultivable waters (an estimated 7,307,642 ha of fresh and brackish water). The potential for fish culture in India has not gone unnoticed, and according to R.V. Pantulu, formerly of the Indian Fresh Water Fisheries Research Institute at Barrackpore, fish culture operations in India have increased sevenfold to eightfold during the last decade. Yet large areas of potentially rich waters still lie fallow and, technologically, Indian fish culture lags behind much of the world.

Though brackish water culture is practised in some areas, notably the states of West Bengal and Kerala, freshwater culture is better developed and will probably remain more important. Freshwater fish culture is favoured not only by the availability of suitable waters and fish, but by the strong preference for fresh fish in most parts of the country, which permits locally raised freshwater fish to compete favourably with saltwater fish in inland areas.

As elsewhere in Asia, the most commonly cultured fish in India and Pakistan are members of the carp family (Cyprinidae). Indian cyprinids used in fish culture are popularly separated into two groups the most desirable species are referred to as major carps; the smallest less desirable species are called minor carps. The minor carps persist in Indian fish culture largely because as fry they are extremely difficult to distinguish from the major carps, and are thus unintentionally stocked. Minor carps may also be deliberately stocked when stocks of major carps are scarce. In the south of India, in the states of Madras and Mysore, a third group, the Cauvery carps, largely supplant the major carps.

Indian carps have not yet become as popular for introduction and culture outside their native habitat as the common carp *(Cyprinus carpio)* or the Chinese carps, but the catla *(Catla catla)* has been introduced to Ceylon, experimentally cultured in Israel and the United States, and commercially raised in Malaysia. All the Indian major carps have been recommended for use in the Philippines.

ECOLOGICAL NICHES OF THE INDIAN CARPS

The Major Carps

The major carps comprise the catla, the rohu *(Labeo rohita)*, and the mrigal *(Cirrhinus mrigala)*. Some authorities also include the calbasu *(Labeo calbasu)*. These three or four species are often grown in polyculture, though their ecological niches are by no means as distinct and well defined as those of the Chinese carps, nor is Indian polyculture nearly as sophisticated as the ancient Chinese practice. In general, the catla feeds on plankton and decayed macro vegetation on the surface and throughout the water column, the rohu is a column feeder on decayed vegetation with a taste for higher plants, the mrigal a bottom-feeding herbivore, and the calbasu a benthic omnivore. Certain of the Chinese carps have been experimentally included in Indian carp polyculture systems and in the future further combinations of Indian carps, Chinese carps, and other fish may be expected in both experimental and practical culture.

The Cauvery Carps

The niches of the Cauvery carps overlap; thus polyculture of these three species could probably be enhanced by introduction of other species. The species association for fish culture in southern

India has already been diversified by the introduction of catla. The catla joins the fringe-lipped carp *(Labeo fimbriatus)*, which consumes mostly filamentous algae and some zooplankton; the white carp *(Cirrhinus cirrhosa)*, a plankton feeder with a preference for zooplankton, and the Cauvery carp *(Labeo kontius)*, which competes with the fringe-lipped carp for filamentous algae, but also eats pieces of plants and detritus.

The Minor Carps

Minor carps present in fish ponds vary from region to region. Among those commonly found in one or another part of India are the reba *(Cirrhinus reba)*, which feeds largely on phytoplankton and decayed plants; the nagendram fish *(Osteochilus thomassi)*, a benthic and marginal feeder on filamentous algae and diatoms; and the sandkhol carp *(Thynnichthys sandkhol)*, which consumes mostly phytoplankton plus some zooplankton; also the bata *(Labeo bata)*, the carnatic carp *(Barbus carnaticus)*, *Barbus chola, Labeo boga, Labeo dyoechilus, Labeo gonius, Labeo nandina,* and *Barbus sarana.*

One more Indian cyprinid should be mentioned: the omnivorous copper mahseer *(Barbus hexagonolepis)*. Not exactly a "minor" carp, the copper mahseer attains a maximum length of 90 cm. It is cultured in some parts of India but is not included in the traditional Indian polycultural scheme.

Spawning the Indian Carps

For many years a major obstacle to the development of culture of the Indian carps has been the inability of culturists to consistently breed them in captivity. All species spawn naturally in rivers and will not reproduce in standing water, although the major carps may be spawned in specially constructed reservoirs, or bunds, where there is enough current to approximate river conditions. The copper mahseer will also spawn in running water ponds provided the temperature is about 20°C, or ripe fish may be hand stripped and the eggs fertilized artificially. Little or no effort has been expended on hand stripping or reservoir spawning of the Cauvery carps or minor carps.

Bund Spawning

The technique of bund spawning can be applied only where

the drainage from an extensive catchment area can be accumulated in a natural depression. The low end of such a depression is blocked off by a strong embankment, so that during the rainy season intermittent streams will inundate the depression. Or a dam may be built in the uplands to concentrate runoff in one place, from which it is channelled into the bund. During the dry season a minimum of 1.5 m of water is maintained in a pool perhaps 3000 m^2 in area, which is stocked with catla, rohu, mrigal, and calbasu in a 4 : 2 : 1 : 1 ratio at a sex ratio of 2 males per female; males may be distinguished by the rough dorsal surface of the pectoral fin. The exact number of fish stocked depends on the extent of the spawning area; about 1 fish per 15 m^2 is appropriate. The pool is usually fished just before the start of the monsoon, and the numbers of each species and sex estimated, so that losses can be rectified by stocking.

The spawning area is a flat piece of land on which grass is grown during the dry season so that the water will not be easily muddied when it is flooded. It may comprise the entire bund other than the permanent pool, or it may be just one corner of the bund.

When the rains come and flood the spawning ground an outlet channel in the embankment is opened up so that water circulates through the bund. The breeders then move out of the pool and begin chasing in the shallow areas. Spawning usually occurs at night during the full moon or new moon.

When spawning is completed, the eggs are collected by dragging seines, about 4 m long x 1.5 m deep, made of mosquito netting, over the spawning ground. Unfortunately a large percentage of the eggs is often destroyed by being trampled by the fishermen. The eggs are placed in spawning pits dug as near as possible to the bund and supplied with bund water. These pits may be 1 to 1.3 m long, 0.6 m wide, and 0.3 m deep and accommodate from 100,000 to 300,000 eggs. When the inlet to the pit is closed, the water temperature rises rapidly, accelerating hatching time to 12 to 18 hours, instead of the 36 hours that might be required at lower temperatures.

The rate of hatching in these pits is low due to bacterial decay and lack of aeration in the stagnant water. Some improvement may be effected by suspending a hatching net or "hapa". in a hatching pit. The hapa is rectangular in shape and constructed as a net within a net. The eggs are placed in the inner net, which is made of material just fine enough to hold them. The newly hatched larvae pass through into the outer net of fine cloth and are nursed there until they are 4 to 5 days old, by which time they have become fry.

Artificially Induced Spawning

As mentioned, the technique of bund spawning may be used only in areas where the topography permits. Induced spawning with the aid of pituitary injections is more generally applicable. The first attempts to apply this technique to the Indian carps were made in 1956, but it is only in the last five or six years that consistent success has been achieved. Today all four major carps and a number of other Indian cyprinids are bred in this manner. Success has been found to hinge on the quality of pituitary glands used and the correct dosage. Pituitaries used in induction should be taken from fully mature, ripe, or freshly spawned fish of the same species as the fish being spawned, or a very closely related species. Dosage is variable, depending on the stage of maturity of the breeders. Since best results are obtained only with fully mature fish, only dosages for such fish will be described.

Breeders are usually selected from 1.5-to 5.0-kg, 2-to 4 year-old fish stocked in spawning ponds a few months prior to the breeding season, which usually coincides with the southwest monsoon; then segregated at maturity. Maturity of males is easily determined; fully mature specimens ooze milt when the abdomen is gently pressed. Selection of females is more difficult, but fish with soft, rounded, bulging abdomens and swollen, reddish vents are preferred. A catheter may also be used to assess ripeness.

Intramuscular injections are made on the caudal peduncle or near the shoulder region Females are injected two or three times, while males receive only one injection. The first injection, of females only, consists of 2 to 3 mg of pituitary extract/kg of body weight, after which the sexes remain separated for 6 more hours. Then males are given a dose equal to the first injection administered to females, while the females receive a second dose of 5 to 8 mg/kg.

Following this injection, the fish are placed together in groups of three (two males and one female) in covered breeding hapas, 1.6 to 6.5 m^2 in surface area and 0.9 m deep fixed on bamboo poles in the marginal waters of ponds. Spawning ordinarily takes place within 3 to 6 hours, but if after 10 to 12 hours no spawning occurs, females only are given a third, slightly higher dosage of pituitary extract.

If the water temperature is near optimum (about 27°C for most species), 60 to 100% success may be expected by this method. Once the eggs have hardened, 8 to 10 hours after spawning, they are transferred, in batches of 75,000 to 1,000,000, to 1.6-m^2, 0.9-m deep hatching hapas set in marginal waters of ponds, where they hatch in 15 to 18 hours at 27 to 31°C. Spawners may be sacrificed to obtain their pituitary glands.

As induced spawning of cyprinids becomes more prevalent in India, selective breeding and hybridization of Indian carps can assume greater importance. Experimental hybridization of Indian carps with each other and with Chinese carps has already been done, but the resulting offspring have for the most part been unpromising ♀ if not incapable of survival. One exception is the hybrid catla x ♂-rohu, which combines the wide body of the catla with the small head of the rohu. These traits would give it an advantage at the market wherever consumers do not eat the head of the fish, while at the same time effectively giving the consumer more nourishment for his money. The catla x rohu hybrid is fertile and has produced a healthy F_2 generation.

Despite recent improvements in induced spawning methods, the main source of stock for culture in India and Pakistan continues to be collections of eggs, larvae, fry, and fingerlings taken from rivers.

COLLECTION OF EGGS, LARVAE, FRY, AND FINGERLINGS

Collection and Hatching of Eggs

Eggs are collected only from the Halda River in the Chittagong area of Bangladesh. Drifting fertilized eggs are captured 12 to 14 hours after spawning, which generally takes place within 3 weeks before or after each full moon from April through July. The precise time of peak spawning is determined by test collections at the start of the season. The number of eggs caught in these collections enables the collectors to forecast the peak of spawning so that they can fish about that time or shortly after, when the maximum number of eggs is available.

The net used in egg collecting is simple, consisting of no more than a rectangular piece of mosquito netting 11 to 12 m long and 2.7 m wide with a bamboo pole attached at each end. Such nets are usually operated by two men in a boat stationed at right angles to the current, but they may also be moored in the river or used by men wading.

Captured eggs may be placed in a compartment of the collecting boat for some time, even up to hatching, but this results in high mortality due to congestion and inadequate aeration. Better methods of holding and hatching eggs involve nets suspended from a bamboo framework, bamboo baskets lined with fine cloth and suspended in the river, or hatching pits dug in the river bank and supplied with water through a system of pipes. Suspended nets are more often used as a temporary device to accommodate the eggs immediately upon collection, after which the eggs are transferred to pits or baskets, which are believed to produce healthier larvae. A typical hatching pit was 448 cm long x 244 cm wide x 46 cm deep and could accommodate 120 to 300 kg of eggs (900,000 to 2,200,000 eggs).

The hatching rate in all these traditional devices is usually 25 to 50%. Somewhat better results may be obtained using hatching pits in conjunction with the double net device already described in connection with bund spawning.

Collection and Transport of Larvae and Fry

The most commonly collected life stages of the Indian carps are the larvae and fry, which are taken at various times of year, depending on the locality. Favoured locations for fry collection are along gently sloping banks of rivers where the current is not too strong and at the mouths of small creeks. In such locations special fry-collecting nets are fixed in 1 to 3 m deep water. These nets are funnel shaped, tapering from 2.5 to 3.5 m wide at the mouth to 20 to 25 cm at the cod end, which is kept open by means of a ring. Two lateral wings may be attached at the mouth so as to cover a wide area. A detachable tail piece, or "gamcha," shaped like a monk's hood, is attached at the cod end. The gamcha may be 1 to 2 m long and 40 to 100 cm wide at the rear end. These and other dimensions of fry nets vary widely according to local need. Equally variable is the mesh of the net, which may be of mosquito netting or nearly any available cloth. Experiments conducted by the Allahabad Substation of the Central Inland Fisheries Research Institute have shown that 0.3 cm netting is much more effective than the materials in general use. Finer netting may be called for where there is a fast current and low turbidity.

Fry nets are anchored in the stream bottom by bamboo poles, two each at the mouth and middle of the net, and one at the tail. Usually a battery of 15 to 25 nets, owned and operated by a group of 7 to 12 fishermen, is placed at each collection site. While the nets are in use two fishermen in a boat move from net to net emptying the gamchas every 1/2 to 1 hour or as often as is necessary to prevent crowding of the catch. Captured fry and larvae are passed through a wire or bamboo sieve to separate them from debris, then transported in large earthen basins to rectangular cloth live wells about 2 m long x 1.25 m wide x 0.7 m deep anchored in the river so that the top is just above the water level. The average daily catch from one fry net may be from 300,000 to 750,000 fry and larvae.

The next step is transportation of the fry to market at one of the larger cities where they are sold to owners of nurseries for raising to fingerling size. The classical container for transport is an open earthen vessel of variable size, called a *"hundi"*. The density of fry in a hundi varies according to the time to be spent in transit.

Table 11.1 Numbers of Indian Carp Fry Recommended for Transport in Earthen Hundies of 27.3 Litre Capacity

Length of fry (MM)	*No. per Hundi*	*Maximum Duration of Transport (Hours)*	*Percentage of Mortality*
12-19	1,500	24	2-5
	1,200	36	2-5
19-25	1,000	20	2-5
	800	30	2-5
25-51	500-800	24	10.0
51-77	200	8	10.0

Table 11.2 Numbers of Indian Carp Fry Recommended for Transport in Closed 11.2 Liter Metal Containers with and without Oxygenation

Initial dissolved oxygen content (ppm)	*Size of fry to be transported (mm)*	*No. of fry to be put in each container*	*Approximate safe period during which transport can be effected (min)*
4	6-7	50,000	19
4	6-7	30,000	31
4	6-7	20,000	47
4	15-20	1,000	40
4	15-20	500	80
4	30	300	120
4	30	150	240
5	6-7	50,000	25
5	6-7	30,000	42
5	6-7	20,000	62
5	15-20	1,000	60
5	15-20	500	120
5	30	300	165
5	30	150	330
6	6-7	50,000	31
6	6-7	30,000	51
6	6-7	20,000	77
6	15-20	1,000	75
6	15-20	500	150
6	30	300	207
6	30	150	414
oxygenated	12-20	1,000-1,250	12 hours
oxygenated	39-51	500	24 hours
oxygenated	25-32	400	16 hours

However many fry are to be transported, it is advisable that the water in the hundies be kept constantly agitated. Manual agitation is sometimes necessary, as when fry are shipped by rail and the train is stopped in a station. Oxygen depletion is further prevented by the addition of 50 to 100 g of colloidal earth to each hundi. It has been suggested that the colloidal earth serves as a buffer to maintain proper *pH,* but its principal function is to attract and concentrate in the bottom sludge dead fry and other potential sources of organic pollutants.

In recent years a variety of more sophisticated transport containers made of metal or plastic and equipped with circulating pumps, aerating devices, or a supply of oxygen have started to supplant the earthen hundi, but many fry are still shipped with no means of oxygenation other than agitation of the water by hand or by the motion of the transport conveyance. Even without supplementary oxygenation, metal containers are sometimes preferred to hundis since there is no danger of breakage. The number of fry which can be transported in closed metal containers with and without oxygen is shown in Table 11.2. As a more general rule of thumb in transporting Indian carp fry, the minimum volume of water per fish has been estimated for different sizes of fry.

On arrival at the market the fry are transferred to 70 to 90 liter earthen pots, or "galmas", each containing 150 to 200 g of colloidal earth. After 10 to 15 min. the colloidal earth is removed and with it the dead larvae and fry.

Young fry sold at market are usually a mixture of major and minor carps, for no method of fry sorting comparable to that used for Chinese carps has been developed for Indian carps, although numerous attempts have been made. Catla can be separated from other species with fair success by placing the mixed fry in a tall, narrow container and allowing the oxygen supply to become severely depleted, at which point the catla come to the surface and most of them can be skimmed off, but the other species do not sort themselves out as the Chinese carps do. About all that can be done otherwise is to separate and discard young predatory fish, most of which are slightly larger than the carp fry, by sieving. Visual identification of advanced fry is possible but excessively tedious.

Table 11.3 Minimum Volume of Water Required During Transport by Indian Carp Fry of Different Size Groups

Length of fry (CM)	*Av. Wt. of fry (g)*	*Minimum water volume required per fry (cc)*
4-7	1.91	25
3-5	0.92	15
2-4	0.35	8
2-4	0.25	7
1-2	0.076	2

Collection and Conditioning of Fingerlings

Fingerlings as well as fry of catla, carnatic carp, and Cauvery carps are collected in Madras and certain parts of East Bengal, Delhi, and Uttar Pradesh. Fingerlings are collected from back waters of rivers, paddy fields, irrigation channels, tanks, and so on, which they enter from large rivers. In such confined waters they may be captured with seines, dip nets, cast nets, or, in Madras, with special rectangular basket traps made of palmyra roots, called "mavulu." These traps, 1.2 m long, 30.5 cm wide, and 0.9 to 1.2 m high, are placed in gaps in dams constructed across channels for the express purpose of obstructing the movements of young fish. A number of holes about 15 cm in diameter are located along the side of the mavulu near the bottom. Fish enter the holes and swim into the trap but find it very difficult to leave due to inwardly narrowing funnel-shaped structures connected to each hole. In parts of Andhra Pradesh state, similar traps are used to collect fingerlings of catla, mrigal, and fringe-lipped carp from irrigated paddy fields. Fry of these species enter paddy fields naturally in June and July and are allowed to grow there until September or October, when they are trapped.

In the Cauvery River delta Indian carp fingerlings are captured by regulating the flow of water through irrigation sluices. Fingerlings congregate below the sluices and are stranded when the sluices gates are closed. This operation is carried out at intervals of about 4 hours and the stranded fingerlings easily captured by seining. Since by the fingerling stage Indian carps have acquired adult specific characters, they are easily sorted visually.

If the fingerlings are to be stocked nearby, they may be transported in jars and stocked immediately, but if the rearing ponds are distant, it is considered advisable to "condition" them so as to eliminate all food and excreta in the gut. It fingerlings are transported in crowded containers without conditioning, feces and vomited remnants of food may severely pollute the water. Conditioning basically involves nothing more than starvation for 48 to 72 hours, although it may also help accustom the fish to crowded conditions. Rapid elimination of food and feces may be achieved by placing the fingerlings in a net basket fixed in a pond, then splashing water on them from all sides, which frightens them so that they pass excreta and vomit immediately. No analysis has been made of the stress factor induced by this practice. It has been suggested that some feeding with animalcules such as cladocerans may be preferable to total starvation in both conditioning and transport.

Practically any container with mesh or perforated sides which can be placed in a pond or stream may be used for conditioning. Conditioning containers should be kept in a shaded area to protect the fingerlings from sudden changes in temperature; an optimum temperature for conditioning is considered to be between 26 and 29°C. Conditioning containers should not be placed where the water will be muddied by the activities of fishermen.

NURSERY PONDS

Preparation and Stocking

Captured and conditioned fingerlings may be placed directly into growing ponds to fatten them for consumption or stocked in rearing ponds for further intensive care. Fry, however, must be nursed for 12 to 15 days to insure their health and satisfactory growth. Both seasonal and perennial ponds, ranging from 7 m^2 to 0.5 ha in area and 0.9 to 3.6 m deep, are used as nurseries. Sample stocking rates for nursing ponds are shown in Table 11.4.

Perennial ponds have the serious disadvantage of harbouring a host of predators, parasites, and competitors of carp fry, which seldom can be entirely eliminated. For this reason, in stocking perennial ponds it is best to scatter the fish, rather than dumping them all in at once. Stocking at night is also advisable, since most fry predators feed visually and the fry are especially vulnerable to predation during the period of acclimatization to a new environment.

Temporary ponds are somewhat better, but the ideal is a pond which can be drained at the discretion of the culturist. In the relatively few modernized nursery units in India, the ponds are drained and sun-dried before use, the bottom is plowed, and a short-season crop of leguminous plants is grown as a source of nitrogen for the soil. After harvesting the legume crop, the plants are plowed under and the bottom is leveled. Further fertilization may be carried out using various kinds of manure or refuse, mixed with oil cake, applied at 200 to 325 kg/ha before the pond is filled. Vegetable manures are usually applied in heaps, weighted down with stones rather than scattered about the pond, so as to reduce the danger of widespread deoxygenation. Animal manure is placed in bags or baskets for the same reason. Dilute sewage is used as a fertilizer in some areas, notably in the Bidyadhuri Spill in Calcutta. In using sewage, care should be taken that the dissolved oxygen concentration in the pond does not drop below 3 ppm. Inorganic fertilizers, though too expensive for use by many culturists, have been used on an experimental basis. They would appear to be especially promising for use in arid areas, where organic manures are relatively scarce and more appropriately used as a source of humus for the soil. Unfortunately, most of the research on inorganic pond fertilization in India has involved use of the N-P-K mixtures, with little testing of the individual elements. Use of N-P-K mixtures may be wasteful, as limnological data from many parts of India show an abundance of potassium. The nitrogen-fixing blue-green algae so prevalent in Indian fish ponds, along with the occasional practice of growing legumes in dry ponds, may make the addition of nitrogen superfluous as well. There is some experimental evidence that phosphates alone are as effective as or better than N-P-K mixtures in many situations, but the lack of adequate controls in this research leaves room for doubt. In parts of Bangladesh a 3 : 1 mixture of cow dung and superphosphate is applied to ponds at 555 kg/(ha) (year). Clearly, pond fertilization practices in India and Pakistan can be improved, through research and by greater appreciation of the fact that the waters of the region vary widely in chemical characteristics, so that each body should be treated

Table 11.4 Sample Stocking Rates of Indian Carps in Nursery Ponds

Type and/or area at pond	*Approximate Stage and/ or size*	*Duration number per ha at water surface*	*at rearing (Days)*
Shallow, seasonal nurseries, 91-152 cm deep, 1,524 cm x 1,524 cm or 1,828.8 cm x 914 cm in area, and paddy fields with a depth of over 30 cm	Fry up to 8.5 mm	741300-1,235,500 depending on density of plankton available as food	15-30
Seasonal nurseries 1,524-1,828 cm x 914-1,219 cm x 91-122 cm.	Fry up to 8.5 mm	1,235,500-1,976,800	15
Seasonal, shallow nurseries 1,524 cm x 1,524 cm in size and 91-122 cm depth	Fry few hours to 3 days old	222,390 without feeding, 1,235,500 with artificial feeding	15
0.62 ha in size	-	-	30
Seasonal nurseries, 0.9-1.5 m deep, 2.4 x 3 m in size, or perennial ponds 1.8-3.6 m deep and 0.5 ha in area	Fry	7,812,500	12-1
Cement cisterns	Fry up to 19-25 mm	8,401,400 without artificial feeding (must be thinned after 10 days)	

individually rather than according to some custom or general formula.

If the soil is acid, it may also be necessary before filling to treat the pond with lime until a *pH* of 8 or 9 is reached. Large perennial ponds are generally acid, but 300 to 500 kg/ha of lime is usually sufficient to correct this situation.

In ponds which cannot be drained and dried, pest control is a major preoccupation of the culturist. Predators include fish such as the snakeheads *(Ophicephalus* spp.), frogs, several varieties of insect, and aquatic birds. Competitors of the carp fry include many species of small, commercially undesirable fish as well as tadpoles. Fishing is usually inadequate to control predator and competitor fish, and poisoning must often be relied on. A number of pesticides of plant origin are used, but the most common is derris root powder. Applied at 4 to 6 ppm, it eliminates virtually all fish as well as killing some aquatic insects and tadpoles. Larger doses may be more effective, but the prescribed dosage is preferred since it initially only stuns the fish, and edible predators can be salvaged by quickly transferring them to clear water. Further control of insects can be achieved by applying an emulsion of 56 kg of mustard or coconut oil and 18 kg of washing soap per hectare. These poisons lose their effectiveness within 2 to 12 days of application, at which time manuring or stocking may be initiated. Ducks and geese must be fenced out of nursery ponds.

Weeds may also cause a problem by competing for nutrients with plankton and by providing hiding places for predators. Manual or mechanical removal is preferable to chemical control, but the possibility of ecological control using shading or herbivorous fishes, should be considered. A lengthier treatment of weed control methods follows in the discussion of growing ponds.

Fertilization of Nursery Ponds and Supplementary Feeding of Fry

The fry remain in the nursery 12 to 15 days and are not ordinarily fed, since plankton produced by fertilization is adequate for growth. It has been customary to stock nurseries when the water turns bottle green, indicating a heavy growth of phytoplankton. Inorganic fertilizers are particularly effective in producing this condition. Recent research, however, has shown that early fry of the major carps prefer zooplankton. Application of fresh or partially dried cow dung at about 11,000 to 16,000 kg/ha will, after about 10 days, produce a vigorous growth of zooplankton lasting 7 to 10 days. Heavier doses are actually less effective in producing zooplankton. This sort of manuring should be carried out before stocking. If the initial dose does not produce an adequate amount of zooplankton, treatment may have to be repeated at 2500 kg/ ha or less every 4 to 5 days until production is satisfactory. Fertilizers applied after the pond is filled should be handled in much the same manner as fertilizers placed in dry ponds; they should be placed in a heap at one corner of the pond. Or a separate culture of zooplankton may be grown in a small, heavily manured pond.

The presence of adequate quantities of zooplankton can be tested for very simply. If 50 litres of pond water are filtered through a plankton net into a test tube and a few drops of formalin or a pinch of common salt is added to kill the plankton, a layer of sediment about 1 cm deep should develop. The colour of the sediment, brown or green, indicates the relative proportions of zooplankton and phytoplankton, respectively.

If a pond simply does not produce enough food organisms, or if other problems such as deoxygenation or reinvasion of predators develop, the fry may be transferred to another pond. This is not necessarily an emergency measure, as a change of environment may accelerate growth in any case.

Table 11.5 Feeding Schedule for Indian Carp Fry

	Artificial food Totalling
First five days after stocking	One to two times the weight of the fry at stocking daily
Second five days after stocking	Two to three times the weight of the fry at stocking daily
Third five days after stocking	Three to four times the weight of the fry at stocking daily

If a heavy bloom of phytoplankton develops, it may be controlled by dissolving cow dung or dye in the surface water to block the penetration of light into deeper water. An abundance of duckweed or other small floating plants serves the same purpose.

Some Indian carp culturists make use of artificial fry feeding.

Various types of dried and powdered oil cakes mixed with rice bran are the customary food, although water fleas *(Daphnia)* have also been recommended. The normal feeding schedule for oil cakes and rice bran is given in Table 11.5.

There has been speculation that oil cakes and rice bran placed in ponds function more as fertilizer than as food. Extensive feeding experiments carried out at the Central Inland Fisheries Research Substation, Cuttack, India, compared the effects on common carp and rohu fry of 19 food items, fed singly or in combinations of two, three, and four items, with the effects of the oil cake-rice bran diet. In laboratory tests many of the diets studied produced better growth 'and/ or survival than did the normal diet. The best of the diets tested, a mixture of notonectids (aquatic insects), prawns, and cowpeas, was then field tested on nursery fry of catla, rohu, and silver carp. Although fry of these species are plankton feeders in nature, all showed better growth and survival on the mixture than on conventional diets. Mrigal fry also grew well on the mixture but were not compared to mrigal fry fed oil cakes and rice bran.

The Biometry Research Institute of the Indian Statistical Institute has been the site of other research on diets for Indian carp fry, in this case involving micronutrients. Addition of yeast, vitamin B complex, and ruminant stomach extract with cobalt nitrate to Fords containing 3-to 26 day-old, *Daphnia-fed* fry of catla and rohu resulted in higher survival and, especially in the case of yeast, better growth. Yeast and B complex also reduced density effects on survival. Yeast, in particular, may find application in commercial culture of the Indian carps. However, while addition of yeast to fry diets enhances growth, it also decreases the total protein per gram dry weight of fry.

Rearing Ponds

When the nursing period is over, at which time the fry have reached a length of 20 to 50 cm, they should be left in the nursery pond for about 2 days, during which time they are not fed, then captured in a fine-mesh seine and transferred to rearing ponds. The nursery pond may then be prepared for the next lot of fry.

Rearing ponds are poisoned, fertilized, and if necessary limed in the same manner as nursery ponds. Stocking rates for rearing ponds are shown in Table 11.6.

Table 11.6 Sample Stocking Rates of Indian Carps in Rearing Ponds

Type and/or area at pond	*Stage and/ or size (MM)*	*Approximate number per ha at water surface*	*Duration at rearing (Days)*
Perennial or seasonal rearing ponds, 122-183 cm deep, 0.62 1.24 ha in area, and paddy fields with a depth of 46 cm or more	25-51	49,420-74,130 without feeding; 148,260-197,680 with regular feeding	30-60
Rearing ponds slightly larger than nursery ponds	25-38	98,840-123,550 without feeding. 148,260-197,680 with artificial feeding	60
Perennial or seasonal rearing ponds retaining water for a long period, but not deeper than 183 cm, long and narrow in shape for easy, inexpensive fishing operations, and paddy fields 45-61 cm deep	19-25	24,710 without feeding; 197,680 with artificial feeding	60-90
Rearing ponds	25-51	4,000-5,000	60
Rearing ponds	35	250,000-500,000	

From this time on, artificial feeding is only rarely practiced in the culture of the Indian carps, the plankton production of a fertilized pond providing adequate food for growth.

PRODUCTION PONDS

Preparation

Production ponds are prepared similarly to nursing and rearing ponds. First all bottom deposits are removed, either manually after draining or, if draining is impossible, by use of nets or long-handled scoops. Lime is added not only to adjust the *pH* of acidic ponds but also as a disinfectant. If a pond has not previously been limed, a heavy dose, perhaps as much as 10,000 kg/ha, may be necessary, but in regularly limed ponds 100 to 200 kg/ha is sufficient, except where the soil is very acidic or poor in carbonates.

Manuring practices vary greatly; one recommended dosage for production ponds is 1000 kg or more of cow dung, 560 to 1200 kg of poultry manure, and 5000 kg of green compost/ha. Other fertilizers such as oil cakes or commercial inorganic fertilizers are also used. Water containing sewage may be used in fattening Indian carps for consumption. This practice might seem questionable from a public health standpoint, but at concentrations of sewage great enough that human pathogens are abundant, dissolved oxygen levels are too low for high survival of fishes.

Table 11.7 Sample Stocking Rates of Indian Carps in Production Ponds

Size stocked	*Fish/ha*	*Species ratio*
50-100 mm	4,000	catla : rohu : mrigal, 6 : 3 : 1
75-130 mm	ca. 11,000	catla : rohu : mrigal, 2 : 3 : 4
juveniles	–	catla : rohu : mrigal, 3 : 3 : 4
80-130 mm	6,250	catla : rohu : mrigal, 3 : 6 : 1
		or
		catla : rohu : mrigal : calbasu 3 : 5 : 1 : 1
–	–	catla : rohu : mrigal, 3 : 3 : 4
		or
		catla : rohu : mrigal :. calbasu 3 : 3 : 3 : 1

Stocking

Stocking practices in Indian carp production ponds seem haphazard when compared with polyculture practices in China and elsewhere. This is partly due to the extreme difficulty of distinguishing the various species as fry, but also because the whole question of suitable polycultural techniques for India has not been adequately explored. Not only are the numbers and proportions of the different species of fish stocked not standardized, neither is the size at introduction to the growing pond. All sizes from fry to juveniles 300 mm long are used. Table 11.7 includes a sample of various stocking policies which have proven at least partially satisfactory.

Since the characteristics of the bodies of water in which these stocking practices are employed are unknown, it is not possible here to make recommendations for pond stocking. In general, the numbers of the major carps stocked should be determined by the relative availability of preferred foods for each. Calbasu are usually added to the basic three species when there are mollusks available in the growing pond, since these are not utilized by the other major carps.

Growth and Yields of Indian Carps

In stocking production ponds, allowance is usually made for an annual mortality, of 30% or more. Very large fish bring poor prices, so Indian carps are seldom left in growing ponds longer than 3 years. Most fish are sold after the first year, at which time catla, rohu, mrigal, and calbasu may have attained weights of 900 to 4100, 675 to 900, 675 to 1800, and 450 g, respectively. The Cauvery carps are smaller fish, attaining first year weights of up to 450 g for the fringe-lipped carp, 330 g for the white carp, and 300 g for the Cauvery carp. In very fertile waters, comparable weights may be reached in as little as 6 to 8 months. Few data are available on yields of Indian carp culture, but they vary widely. In semiwild waters, where the fish are merely stocked and forgotten until harvest, yields seldom exceed 110 kg/ha. With cultivation this expectation may be increased to 300 to 900 kg/ha. With artificial feeding yields as high as 2802 kg/ha have been achieved. In recent experiments at the Central Indian Fisheries Research Station at Cuttack, the unprecedented yield of 3564 kg/ha was obtained by stocking catla, rohu, mrigal, silver carp *(Hypophthalmichthys molitrix)*, grass carp

(Ctenopharyngodon idellus), common carp, and calbasu in the ratio of 5 : 10 : 5 : 10 : 4 : 5 : 1 at 5000/ha, with fertilization and supplementary feeding.

Marketing

Indian carps usually are sold fresh, in accordance with the regional preference. Most fish are not sold directly to the consumer, but through various middlemen; sometimes as many as five are involved in handling one lot of fish. Obviously, this greatly increases the price to the consumer and reduces the amount of protein food available to those who most need it. In some states of India, notably Ahmedabad and Maharashtra, cooperatives market a substantial proportion of fishery products at considerable saving to the consumer, a practice which may be extended to aquacultural products.

Problems of Indian Carp Culture

Among the routine problems facing Indian fish culturists is the presence of unwanted plants in growing ponds. Depending on the plant species to be dealt with, a diversity of methods, including poisoning, is used in weed control. Many of the most effective poisons are chlorinated hydrocarbons such as 2,4–, which are scarcely to be recommended for introduction into ecosystems or human food supplies. Other poisons, such as copper sulfate, sodium arsenite, and anhydrous ammonia gas, lack the long-term cumulative effects and the dangers to man presented by the chlorinated hydrocarbons, but they may be toxic to fish. In all but the most desperate cases therefore, it is safer (and usually cheaper) to use mechanical or ecological methods of plant control. For floating weeds, outright manual removal is the best treatment. Emergent plants may be controlled by cutting of leaves at weekly intervals for about 6 to 8 weeks before fruiting. Marginal weeds may be controlled by many methods, including plowing under, grazing by livestock, burning during the dry season, or deepening the margin of the pond. Rooted submerged weeds may be removed, albeit with considerable labor, by netting with strong nets, by dragging chains, or with the aid of various types of fork and rake.

The commonest agents of ecological weed control are herbivorous fish. The most commonly used species for this purpose is the grass carp, which in addition to destroying or reducing most submerged and emergent plants and adding to pond fish production, contributes to the nourishment of other fish by dropping partially digested plant remains in its feces. All four of the major carps and a number of the minor carps, although unable to utilize fresh macrophytes as food, consume the partially decayed plant material in grass carp feces. However, grass carp are not a panacea for weed problems, since there are some plants, such as *Eichornia* and *Salvinia,* which they eat only with reluctance if at all. There is also the possibility that if grass carp find their way into natural waterways, they may be destructive of the native flora. Another herbivore, already introduced to Ceylon for use in weed control, is the tawes *(Barbus gonionotus).* A number of species of tilapia, including *Tilapia melanopleura, Tilapia mossambica, Tilapia nilotica,* and *Tilapia zillii,* may effectively control certain types of weed. However, *T. mossambica* has on at least one occasion been found to depress total yield in polyculture involving catla.

Another ecological method of weed control is shading. Trees around the border of a pond may provide sufficient shade to discourage marginal weeds, but to control offshore weeds in most ponds the culturist must rely either on floating plants such as duckweed *(Lemna),* individuals of which are small enough not to interfere with netting and routine management operations, or on creation of an algal bloom. A heavy bloom can be induced by repeated application of N-P-K fertilizers, but this may be prohibitively expensive. An unusual combination of effects results from application of superphosphate or urea at 50 ppm or more. These substances at that concentration are toxic to most submerged plants, but they act as fertilizers and produce an algal bloom as well. The desirability of an algal bloom must of course be evaluated in terms of the characteristics of the individual pond and the feeding habits of the fish present.

Other pond management practices used in India and Pakistan include the treatment of foul water by doses of 1.5 ppm or less of potassium permanganate, raking the pond bottom to release accumulated gases, and addition of minute amounts of alum to settle suspended or colloidal matter and reduce turbidity.

Among the diseases reported in Indian carps are gill rot, *Saprolegnia* infection, dye disease, fin rot, Ichthyophthiriasis, costiasis, argulosis, ligulosis, gyrodactylosis, and dropsy. For a detailed discussion of some of these diseases and methods of treatment, the reader is referred to Davis.

12

Culture of Catfishes

Recent progress in the culture of the channel catfish *(Ictalurus punctatus)* in the United States has served to focus attention on the potential for culture of other catfishes. The suborder Siluroidei is large and varied but, in general, its larger members, which are represented in the inland and coastal waters of every continent, are of high quality as food fish and adapt well to heavy stocking and artificial feeding. Many are also extremely hardy with respect to environmental conditions and can be grown where few other fishes will survive. Their only serious drawback from a fish culturist's point of view is their great voracity, which makes supplemental feeding a must. This expense is compensated for somewhat by their catholic taste.

CULTURE OF PANGASIUS SPP. IN ASIA

Capture of Wild Fry and Species Used

Though channel catfish farmers in the south central United States may be the most highly publicized growers of catfish, they were by no means the first. The history of catfish culture probably began on the Indochinese peninsula, where a number of members of the genus *Pangasius* (family Siluridae) have been grown in pond since ancient times. The most important species is *Pangasius sutchi* (pla swai), which reaches a maximum length of 150 cm. The smaller *Pangasius larnaudi* (pla tepo), which reaches 70 cm, is less important, but it is economically significant. *Pangasius micronemus* (pla sangawad), which grows to only 50 cm, is less esteemed as a species for culture, but it is sometimes mistakenly stocked with the two preferred species. *Pangasius sanitwongsei* (pla thopa), considered a good food fish, is also excluded from culture where possible, because it is too large–up to 250 cm.

As is the case with many forms of Asian fish culture, culture of *Pangasius* spp. was originally dependent on the capture and rearing of naturally produced fry, and a substantial proportion of the total production still originates in this way. Natural spawning occurs in large rivers at various times between June and November, and it is then and there, particularly in the Chaophya River, the Sakakrong River, and the Kreang Kri Canal of Thailand, and the Mekong River in Cambodia, that 3-to 8-cm fry are captured in 1-cm mesh seines. Juvenile fish, weighing 80 to 150 g each, are also collected, with dip nets, from the Tonle Sap and Grand Lac of Cambodia during February to April.

Inadvertent stocking of *Pangasius micronemus* is due largely to the difficulty of distinguishing between fry of that species and those of *Pangasius sutchi* and *Pangasius larnaudi.* The identifying morphological and behavioural characteristics of these three species have been worked out by officers of Thailand's Department of Fisheries (Table 12.1). Less is known about the fry of *Pangasius sanitwongsei,* which has a somewhat limited distribution, and is apparently commonly cultured only in Vietnam.

Spawning and Fry Rearing in Captivity

In the 1960s, in Thailand at least, the effects of overfishing, pollution, and destruction of spawning grounds began to be felt in the scarcity, and consequent high price, of *Pangasius* fry. Accordingly, in 1966, Thai fishery authorities began to apply the results of earlier experiments with the induction of spawning of P. *sutchi* by means of pituitary injection. Both sexes receive fractional injections of P. *sutchi* or *Clarias batrachus* pituitary, followed by fertilization by the dry method. Proper dosages and time intervals for injections have yet to be worked out, thus the range of hatching rates is extremely wide, from 0 to 85%.

Immediately following fertilization, the eggs are placed in fine mesh hatching nets containing fine fibrous materials-either aquatic plants or such artificial substitutes as palm and jute fronds-and allowed to attach. Hatching occurs 24 to 33 hours after the nets are placed in clear water at 26.5 to 31.0°C.

The mouths of the larvae open about 36 hours after hatching, and feeding commences about 12 hours later. Zooplankton is the

Table 12.1 Distinguishing Characteristics of Fry of *Pangasius* Spp.

	P. sutchi	*P. larnaudi*	*P. micronemus*
Mixing of shoals	With or without *P. larnaudi*	With or without *P. sutchi*	Without both species
Movement of caudal fin	Fast movement, splashing water at the surface	Faster movement than *P. sutchi*	Slow movement of caudal fin, no water splashing
Ratio of head width. to body length	1 : 8	1 : 6	1:7
Angle of insertion of dorsal fin base	Parallel to body axis	Parallel to body axis	At about 30" to bod y axis
Number of ventral fin rays	8-9	6	6
Mouth cleft	Wide	Wider	Narrow
Number of gill rakers	More than 12 long gill rakers	12 short gill rakers	More than 12 long gill rakers
Dorsal and pectoral fin color	Greyish black	Absolutely black	Colorless
Color of upper and lower lobes of caudal fin	Greyish black	Absolutely black	Colorless

chief natural food of the early fry, but small insect larvae and worms also may be taken. When availability of these foods is poor, cannibalism may occur. Fry are maintained in the hatching tanks for 16 to 21 days, by which time they attain lengths of 2 to 6 cm and may be stocked in rearing ponds. The best survival of very young fry (32.2%) has been achieved by feeding live *Daphnia.* After the fry attain a length of about 2 cm, *Daphnia* may be supplemented with small pieces of mollusks, worms, and cooked fish.

Growing for Market in Ponds

Pangasius spp. are stocked alone in ponds, or in combination with tawes *(Barbus gonionotus)* or sepat siam *(Trichogaster perctoralis).* When stocked at about 25/m^3, and fed on kitchen waste, bananas, cooked broken rice, rice bran, and soft aquatic and terrestrial plants, *Pangasius larnaudi* fingerlings may attain weights of 0.15 kg at the end of 1 year, and 1 kg in 2 years. *Pangasius sutchi* grows more slowly initially, but may reach 4 kg in 2 years. When fish wastes are added to the diet, as is common in Vietnam, weights of 1.0 to 1.2 kg may be reached in 8 to 10 months.

Data on the yields obtained by pond culture of *Pangasius* spp. are lacking but there are verbal reports that far higher yields are obtained through culture in floating cages set in series along the edges of rivers, where there is a gentle current. Very high yields are also obtained when *Pangasius* spp. are used as one component of a pig-poultryfish farming complex.

Growing for Market in Floating Cages

The floating cage technique, which is traditional in Cambodia, has been practiced in Thailand for only a little over 50 years. Fry and fingerlings are initially stocked at 150 to 300/m^3 in cages 1 to 9 m^2 in surface area and no more than 1.5 m deep, made of bamboo poles or wooden planks, covered with mosquito netting. In each cage a food plate is suspended about 10 cm under the water surface. The fry are fed once or twice daily with ground trash fish mixed with boiled rice or rice bran. As soon as they reach 15 cm in length, the young catfish are stocked in larger cages.

Cages used for growing *Pangasius* for market are more commonly built of 25 mm planks than of bamboo, and they may incorporate living quarters for the culturist. A gap of about 25 mm is left between planks to permit passage of water. Growing cages vary from 2 to 16 m long, 1 to 6 m wide, and 1.5 to 2.5 m deep; the most popular size is 4 m x 3 m x 2 m. Normally bamboo poles are used as floats in order that the top of the cage, which is provided with 1 m x 1 m holes so the fish may be fed, remains about 30 cm above the water surface. Though the cages themselves last 10 to 12 years, the bamboo floats must be replaced every 2 years.

High stocking rates (20 to 60 kg/m^2) are employed to offset the usual high mortality in cages. Trash fish is the basic feed during times of year when fish are plentiful; the daily feed ration is 10 to 15% of the weight of catfish. When trash fish are scarce, maize, broken rice, rice bran, and aquatic vegetation in large amounts substitute for or supplement the fish diet.

Pangasius spp. in cages may be reared from 0.08 to 1 kg in 8 to 10 months. If the growing period is extended to 2 years, weights of 2 kg may be reached.

Culture in Laos, Malaysia, and India

Pangasius spp. are not limited in their distribution to Thailand, Cambodia, and Vietnam but range throughout southeast Asia, from East India to the Malay Peninsula, Indonesia, and neighboring islands. Over much of this area they are not cultured, in some regions perhaps because of the lack of suitable grounds for collecting large numbers of fry or fingerlings. This may be the case in Laos, where *P. larnaudi* and *P. sutchi* fingerlings are imported from Thailand in January and stocked at 25/ m^3 in ponds 0.2 ha in area and 3 m deep. Laotian culture techniques are generally the same as those employed in Thailand and Cambodia, but fish are not used as food. Fertilization with pig and chicken dung is carried out, and it is believed that the fish consume some of the resulting plankton. The growing season is limited to 8 months, and growth is relatively poor, with fish averaging only 200 g in weight when harvested.

The only known instances of culture of *Pangasius* spp. outside Indochina occur in Malaysia, where *Pangasius micronemus* may be stocked in combination with lamplai *(Barbus schwanenfeldii)* and sepat siam, and, in the Indian state of West Bengal, where *Pangasius larnaudi* and/or *Pangasius sutchi* are cultured on a limited scale.

Culture of *Clarias* Spp. in Thailand

One of the most widely distributed catfish families is the Clariidae, members of which are found throughout southeast Asia, the Indian subcontinent, and Africa, as well as in parts of the Near

East. The Clariid catfishes are distinguished by the possession of an accessory air breathing organ, which enables them to exist for hours at a time out of the water, or indefinitely in oxygen poor waters and even moist mud. They may take further advantage of their versatility by coming ashore at night in search of food. The extreme hardiness of the Clariids renders them well suited to culture in arid regions such as tropical east Africa, where their use has been promoted in the last decade.

Interest in culture of Clariid catfishes outside of tropical Africa is even more recent, but already spectacular success has been achieved in Thailand, where *Clarias batrachus* and *Clarias macrocephalus,* two of the most highly valued native fish, support a thriving industry which is becoming progressively more efficient C. *macrocephalus* is preferred by consumers for its more tender flesh, but culturists favour C. *batrachus* because it grows faster, and it has become the dominant species.

Collection of Wild Fry

The major source of fry of both species is natural spawning. C. *macrocephalus* spawns during the rainy season in nests constructed on the bottom of paddy fields in 20 to 50 cm of water. C. *batrachus* spawns at a similar depth in a horizontal hole in the bank. Hatching requires 20 hours at 25.0 to 32.2°C.

Fry are collected in hand nets from May to October; generally 2000 to 15,000 fry are collected from a nest. About 50 million fry are collected annually in this manner.

Artificial Propagation

Collection of naturally produced fry, extensive though it is, cannot provide enough stock for commercial culture. Accordingly, Thai fishery officials have undertaken, with considerable success, to spawn *Clarias* spp. in captivity. As early as the mid-1950s it proved possible to spawn C. *batrachus* by methods similar to those used in spawning channel catfish in the United States. In place of the metal or wooden containers used to spawn channel catfish, horizontal holes, 20 to 35 cm in diameter, were dug in the bank a little over 1 m apart, and aquatic plants were provided nearby. About 80% of the pairs stocked near such holes spawned within 7 to 10 days, yielding 2000 to 5500 fry per spawning.

C. *macrocephalus* was induced to spawn by means of pituitary injection. Single intramuscular injections of 100 to 190 g fish with 13 to 26 mg of pituitary extract per kilogram at 25 to 32°C produced spawning within 14 to 16 hours in 60 to 80% of cases, even when the breeders were placed in small aquaria.

Fertilized eggs are placed in shallow troughs and jars, where hatching takes place within 20 hours at 26 to 33°C. The water should be changed twice daily and fungused eggs removed as soon as they are detected.

Fry Rearing

The larvae absorb their yolk sacs after 5 days, at which time they are transferred to earthen ponds about 3m^2 in surface area. Ponds up to 1 m deep may be used for fry rearing, but best results are obtained in water 10 to 18 cm deep. Fry may be stocked very heavily-5000 to 6000/m^2-for the first few weeks of life. The best early food is zooplankton; after 2 or 3 weeks, boiled fish may be added to the diet. When the fry reach 1.5 to 10.0 cm in length, they are ready to be stocked in production ponds.

Growing for Market

Ponds used in growing *Clarias* spp. for market are 100 to 1000 m^2 in area and 1 to 3 m deep. They are prepared for stocking by firming the banks to discourage climbing or burrowing, and by erecting a fence about 50 cm high around each pond. If a small amount of water is allowed to continually run into the middle of the pond at the surface, the fish may be discouraged from digging into the bottom.

Fry in production ponds are stocked at about 180/m^2 and fed on ground trash fish, mixed with rice bran at a ratio of about 9 : 1. For the last 15 days of growth the feed formula is changed to a 17 : 2 : 1 mixture of ground trash fish, broken rice, and rice bran. Fertilization, with about 30 kg of farmyard manure per 100 m^2, is carried out only if growth is poor.

By use of these methods, three crops of C. *batrachus* may be annually grown to marketable size (about 26.5 cm and 145 g). Thai culturists thus achieve an average survival rate of 37%, a 5 to 6 : 1 food conversion ratio, and the tremendous annual yield of 97,000 kg/ha.

CULTURE OF CLARIID, PLOTOSID, AND SILURID CATFISHES

Clariids

C. *batrachus* is grown in cages like *Pangasius* spp., but on a smaller scale. C. *batrachus* is sometimes stocked as a minor component of polyculture systems in ponds and rice fields in India, Pakistan, Malaysia, Veitnam, and the Philippines. Only in the Philippines it is occasionally fed, on trash fish and small shrimp, and in none of these countries it is intensively cultured.

The highly productive culture of *Pangasius* spp. and *Clarias batrachus* in Thailand and Cambodia, along with the recent culture of the channel catfish in the United States, has already sparked interest in culture of catfishes native to other regions. Many countns are experimenting with culture of *Clarias lazera.* They have solved the problems of large-scale spawning and have developed an artificial feed, made of dried liver, dried vegetables, yeast, and other ingredients, which has enabled them to rear substantial numbers of fry past the stage where cannibalism is a serious threat. At present they are capable of producing fair amounts of edible sized fish, but only with high rates of stocking and feeding. Experiments conduced at the Serrow Fish Farm indicate that polyculture of C. *lazera* with common carp and *Tilapia* spp. would result in better yields than are presently achieved through monoculture of carp, the prevailing mode of fish culture in the UAR.

Another member of the Clariidae, *Heteropneustes fossilis,* has been cultured for some time in Pakistan. Indian biologists have recently shown an interest in the species, and have had some success with artificially induced spawning.

Plotosids

Two estuarine species of the Australasian catfish family Plotosidae, *Plotosus anguillaris* and *Plotosus canius,* occur as incidental fishes in Indonesian tambak culture of milkfish *(Chanos chanos)* and another Plotosid, the freshwater *Tandanus tandanus,* is sometimes marketed as a food fish in Australia. There are no established fish culture enterprises of any magnitude in Australia, but when the possibility of freshwater pond culture is suggested, *T. tandanus* always figures in the speculations. Experiments carried out at the Inland Fisheries Research Station, Narrandera, New South

Wales, indicate that this catfish can easily be kept and bred in small ponds. It is a nest building species, and for this purpose gravel should be provided, though it will spawn, with much less success, on a mud bottom. It seems to be necessary that the water temperature be at least 24°C for the fish to engage in reproductive behaviour. Spawning, which has been observed in 0.5 to 1.2 m of water, may be accelerated by a slight rise in the water level, but this is not a necessary condition. A subsequent drop in the water level will cause the abandonment of nests. One or both parent fish guard the nest until after hatching, but it is likely that, as is the case with many species exhibiting similar behaviour in nature, parental care would prove unnecessary in intensive culture, where predators are excluded. It is said that *T. tandanus* can be sexed by observing the shape of the urogenital papilla, but details of sexing are not known to us.

Silurids

Traditional Chinese fish culture methods make no use of catfish, but the native Silurid catfish *Parasilurus asotus* is sometimes included in pond communities in Taiwan, and the possibility of its culture, as well as that of the Bagrid catfish *Pseudobagrus fulvidraco*, is being studied in mainland China.

Another Asian Silurid, *Wallagonia attu*, sometimes called "freshwater shark" because of its high dorsal fin, is sometimes used in low-intensity fish culture in reservoirs and swamps in Pakistan. *W. attu* is not stocked in ponds because of its extremely voracious and piscivorous feeding habits. *Ompok bimaculatus*, which is occasionally stocked in ponds, has been artificially spawned by means of injections of cyprinid pituitary extract.

Culture of Siluris Glanis in Europe

A Silurid catfish which has been more thoroughly studied is the sheatfish *(Siluris glanis)*, native to many of the large rivers of Europe. Sheatfish have occasionally been used as predators to control excess reproduction in pond culture of various fishes in Europe, but it is only in the last few years that they have been bred on a large scale, first in Hungary, then in Yugoslavia and Czechoslovakia.

Artificial Propagation

Sheatfish spawners may be obtained from rivers, or they may

be overwintered in running water in 300-to 800-m^2 storage basins 1.0 to 1.5 m deep. In the latter case, 30 to 40 kg of small fish must be provided monthly as food for each 100 kg of sheatfish. The storage basins may double as spawning ponds, but the fish must be segregated by sex in the spring, before the spawning season, as soon as the water temperature reaches 12 to 14°C. The sexes may be distinguished externally by the shape of the genital papilla, which is narrow, pointed, and nipplelike in males and wide and rounded in females.

When the water temperature reaches 20°C, the basins to be used for spawning are filled to a depth of 1.0 to 1.2 m, and artificial spawning nests are constructed. The first step in constructing a sheatfish spawning nest is to bind three 1.8-to 2.0-m sticks together to form a tripod, whose legs are inserted in the pond bottom. Roots of willow or alder or twigs or pine or thuya are suspended from these frames to collect the adhesive eggs. Mats made of reeds are placed under the collectors to salvage eggs that fail to attach. The number of nests placed in each spawning basin, usually 2 to 4, corresponds to the number of pairs of sheatfish stocked. The spawners are placed in the basin the same day it is filled, and spawning usually occurs within a few days, if not the following night.

The nests are checked daily from a small boat. Those that contain eggs are removed and the twigs and mats, with eggs attached, are hung on wires and suspended in covered hatching crates 150 cm x 50 cm x 60 cm deep, made of 0.5-to 0.8-mm mesh perlon. The loaded crates, each containing 30,000 to 50,000 eggs, are affixed next to the inflow pipe in storage basins similar to those used in overwintering and spawning. Hatching occurs in about 60 hours at 20 to 22°C. The population density of the newly hatched fry is adjusted to 50 to 70/m^2. Basins containing young fry are provided with ample cover in the form of twigs, reeds, and bricks.

Rearing Fry and Growing for Market

When, 4 to 5 days after hatching, the fry attain the length of 12 mm, they begin feeding. At this time it is necessary to provide large amounts of plankton. At the age of 10 to 12 days the fry may be switched to a diet of ground fish and scrapings of liver and spleen. These foods are presented by smearing them on a black tray or the side of a flowerpot. In addition to provision of shelter and heavy feeding, periodic removal of the fastest growing fish is necessary to forestall cannibalism.

At the age of 30 to 35 days the 5-to 6-cm sheatfish fingerlings are stocked in ponds at 1500 to 2000/ha. A ready supply of food can be assured by spawning common carp *(Cyprinus carpio)* and tench *(Tinca tinca)* in the pond 15 days before stocking it with sheatfish. At the end of their first summer, the young sheatfish weigh 50 to 100 g and are ready for stocking in carp ponds at 50 to 150/ha. Most sheatfish are harvested along with the carp at the end of the following summer, at which time they weigh 0.5 to 1.4 kg. In some cases enough 2-year-olds are left to populate the pond at 35 to 60/ha. These fish are harvested the following fall as 1.5-to 2.0-kg specimens.

Worldwide Prospectus of Catfish Culture

As the success of Asian catfish culturists receives more publicity, it is certain that the transplantation of some of the Asian catfish in other areas will be suggested. One accidental introduction has already occurred in Florida, where *Clarias batrachus,* the albino form of which is commonly brought into the United States as an aquarium fish, has become acclimated. Ecologists and sport fishermen tend to view the "walking catfish" as a destructive force in the Florida ecosystem and hope to limit its spread. Local fish culturists, who have on occasion seen their channel catfish undersold by imported catfishes, also have a negative opinion of C. *batrachus.* However, recently Fred Meyer of the United States Fish and Wildlife Service has admitted that "Exotic fish such as *Clarias* and *Pangasius,* African and Asian catfishes, though in disrepute at the moment, could conceivably replace the channel catfish."

The fears of sportsmen and ecologists seem well grounded, and it is recommended that introductions of exotic catfishes, in the United States and elsewhere, be undertaken cautiously, if at all. On the other hand, there is no reason not to follow the example of the Hungarian sheatfish culturists and proceed energetically in exploring the possibilities of culture of native catfishes throughout the world.

13

Frog Culture

FROG FARMS

The status of commercial frog culture is nebulous. On the one hand, government publications repeatedly advise prospective frog farmers that intensive commercial culture of frogs as food animals has yet to be achieved. On the other hand, one continually encounters individuals who claim to be operating profitable "frog farms." Such establishments generally turn out to be slightly modified shallow ponds or swamps, where frogs are harvested in much the same manner as wild frogs. In some cases, husbandry is limited to erecting a fence to retain the frogs and exclude predators, and the chief market is other would-be frog farmers. Other such farms, however, are more sophisticated and sell to restaurants and other food outlets. Intensive indoor culture methods have also been developed for several species, but at present they are applied only to the production of laboratory frogs, and opinions differ as to whether modifications of these methods could economically be applied to the culture of frogs for human consumption.

Source of Stock

Most attempts at frog culture have been made in the United States, where frogs are among the most expensive luxury foods. Numerous species are harvested from the wild and are generally not discriminated among by buyers or consumers except on the basis of size. The largest and the most widely used in attempts at culture is the bullfrog *(Rana catesbiana)*. Bullfrogs lay their eggs in shallow standing water during April in the South and May or June in the North. Hatching requires 4 days to 3 weeks, depending on temperature. The aquatic larvae, generally known as tadpoles, feed chiefly on benthic algae. In 5 months to 2 years they metamorphose into the semiaquatic and exclusively carnivorous adults, which may reach lengths of up to 20 cm.

Prospective frog farmers may obtain bullfrog stock from commercial sources, or eggs or tadpoles may be taken from the wild. Bullfrog eggs (and those of a few other large frogs) may be distinguished from those of small, undesirable species by the size of the floating egg mass, which covers about 0.5 m^2. size is also the distinguishing characteristic of bullfrog tadpoles, which are much larger than most other tadpoles of the same age.

Stocking

Whichever life form is taken, they should be distributed around the perimeter of the body of water to be stocked. Although this method may be suitable for new operations, the culturist should endeavour to breed his own stock as soon as possible. In addition to the taxonomic uncertainty in collecting wild stock, wild tadpoles often harbour pathogenic organisms and suffer high mortalities in the late stages of metamorphosis.

All known attempts at frog farming in the United States have failed to incorporate any form of control over breeding but have allowed the frogs to spawn as they would in the wild. The resulting tadpoles have sometimes been reared separately from adult frogs, and adults have been segregated by size in the belief that extensive cannibalism would otherwise result. However, some researchers have found cannibalism rare or nonexistent among well-fed bullfrogs of all ages.

Feeding

Feeding is the key not only to averting cannibalism, but to health and satisfactory growth of frogs. Frogs and tadpoles maintained outdoors will obtain some food naturally, but at commercially feasible population densities, the natural food supply must be supplemented.

Tadpoles, while primarily herbivorous in nature, will accept any soft animal or vegetable matter. Among the feeds which have been employed are boiled potatoes, meat scraps, and chicken viscera. An especially attractive idea is the use of the viscera and other scraps from butchered frogs as tadpole feed.

Once metamorphosis to the frog stage is complete, feeding becomes much more difficult. Adult frogs feed exclusively on moving animals. Japanese researchers have reportedly been able to induce frogs to ingest stationary silkworm pupae by mixing them with live nightcrawlers, then removing the nightcrawlers when the frogs have become accustomed to the pupae. Another Japanese method of eliminating dependence on live food involves the use of wooden trays containing about 12 mm of water. Dead silkworms or other food items are placed in the trays, which are anchored in shallow water near shore. A small motor keeps the trays oscillating slowly, so that the silkworms roll back and forth.

So far as is known, these Japanese techniques have not been tried with bullfrogs in the United States. Most American frog culturists have relied on stocking or attracting live food animals. One farmer, located near the ocean in Florida, stocked his ponds with marine fiddler crabs (*Uca* spp.), which are abundant on Florida beaches. Smaller species of frogs and their tadpoles may be eaten by bullfrogs and are sometimes stocked as food. Aquatic plants may be encouraged as shelter and food for tadpoles, crayfish, and other potential food animals. Terrestrial flowering plants serve a similar function by attracting flying insects, which are hunted by frogs on shore. More insects may be attracted by illuminating the shore of the pond at night with 100 to 200 W clear lamps. No combination of these methods has yet proven entirely satisfactory, and the difficulty of supplying adequate amounts of food remains the principal obstacle to successful bullfrog farming.

Design of Ponds

Another problem faced, and largely overcome, by early experimental frog culturists is territoriality. A large bullfrog may require about 7.5 m of shoreline as a feeding territory, a trait which severely limits the number of bullfrogs in most natural environments. Natural bodies of water, however, usually include large expanses of deep, open water which are of little use to bullfrogs during the growing season. Culturists are therefore able to maintain frogs at population densities much higher than those usually observed in nature by reducing the amount of open water in their ponds, while increasing the length and irregularity of the shoreline through construction of islands and peninsulas extending into the centers of the ponds. If natural ponds suitable for such modification are not available, the shoreline of artificial ponds may be maximized by constructing them as a series of narrow trenches. Such trenches should run north and south insofar as possible, so that vegetation on the banks will serve as shade for the frogs.

Whatever form of pond is used, a small portion of it should be deep enough to protect frogs and tadpoles from extreme heat or cold. In the South, 30 to 45 cm is adequate, but in North, deeper water may be necessary to insure the survival of hibernating frogs. In any location, a large portion of the pond should be only 5 to 15 cm deep to facilitate the feeding behaviour of frogs and tadpoles.

Predators of frogs and tadpoles are numerous, and some tadpole predators, such as large aquatic insect larvae, are virtually impossible to exclude from frog ponds. Some terrestrial predators may be kept out by enclosing ponds with a small-mesh wire fence about 1 m high, sloping outward at an angle of 35°. Birds are more difficult to exclude, but a wire net stretched above the shallows may be partially effective. No predator control method developed to date is 100% effective, and the culturist should allow for some loss.

Harvesting

Harvesting of bullfrogs from ponds as just described is extremely inefficient. The methods employed are the same used in hunting wild frogs-fishing with hook and line, spearing, and hand capture. Live bait is occasionally used in fishing for frogs, but a more common practice is to dangle a crude lure, made of red cloth or yarn, in front of the frog to simulate a hovering insect. Spearing and hand capture are done at night with the aid of a bright spotlight, which dazes and immobilizes the frogs. Clearly, these methods must be supplanted by mass harvesting techniques if successful commercial pond culture of frogs is to become a reality.

Data on the commercial status of frog culture in the United States are few and sometimes contradictory. Most of the reports of success have come from Florida and Louisiana, where the growing season is very long, if not year-round, but commercial suppliers of frogs are located as far north as Vermont and Wisconsien.

INDOOR CULTURE OF FROGS

Species Cultured

A newer approach to frog farming, and one which largely eliminates climatic considerations, is indoor culture. Practical methods of indoor propagation and rearing of frogs for use as laboratory animals were first worked out by T. Kawamura of the University of Hiroshima, Japan, Kawamura's methods were subsequently adapted for use with American species by George W. Nace of the Department of Zoology, University of Michigan. Nace's methods have been the model for several other institutions which have undertaken to produce their own experimental animals, but to date no one has attempted indoor commercial culture of frogs as food animals. However, the species routinely cultured at Michigan include the green frog *(Rana clamitans)*, the pickerel frog *(Rana palustris)*, and the leopard frog *(Rana pipiens)*, all commonly marketed as food, and there is no reason to believe that any of the larger American frogs could not be similarly cultured. Thus the possibility of indoor commercial production of frogs for human consumption cannot be ignored.

Water Supply

As might be expected in an indoor culture system, pains had to be taken to provide a suitable water supply for the University of Michigan Amphibian Facility. Based on the experience of Nace and his associates, there are four requirements for maintenance of a self-perpetuating frog colony:

1. The water supply must be abundant at all times; the Michigan facility uses up to 90 litres/min.
2. Line pressure should be adequate to permit individual fine control of flow through each container.
3. The *p*H should be slightly acid.
4. The water temperature must be constant at 20 to 22°C.

The source of water at Michigan is the city of Ann Arbor municipal system, which is unsatisfactory on the last three counts. Booster pumps, pressure regulating valves, and carefully designed plumbing have been installed to compensate for irregular main pressure, *pH* is maintained at 6 to 7 by the monitored introduction of acetic acid, and the temperature is regulated with the aid of heaters and industrial capacity mixing valves. A commercial culturist would of course seek to locate so as to avoid these expenditures if possible.

Ann Arbor city water is chlorinated, which would seem to present yet another problem, as it does for tadpoles. It has proved necessary to install an industrial activated charcoal dechlorinator to provide safe water for tadpoles. However, while chlorine is toxic to tadpoles at concentrations well below the 0.6 ppm found in Ann Arbor city water, adult frogs can stand chlorination up to 4 ppm. In fact, mild chlorination serves as a prophylactic measure against bacterial diseases. Adults are thus kept in water provided by lines which bypass the dechlorinator.

Tadpole Bottles

Differences in water quality requirements, along with other aspects of the life history of frogs, dictate that separate housing facilities be provided for each life stage. Fertilized eggs and very young tadpoles are held at low density in shallow enamel pans. Dead embryos are removed regularly, and the water is changed at least every third day. When vigorous swimming commences, the tadpoles are transferred to special tadpole bottles.

A tadpole bottle is constructed by removing the bottom of a conventional I-gal glass or plastic bottle, stoppering and inverting it. Water is supplied from below through a 10 mm glass tube and removed through a 15-mm plastic siphon tube extending to the stopper. In this manner a constant flow is maintained, and the oldest, "stale" water is removed. Flow through each bottle is adjusted so that the water is exchanged about three times daily, yet dangerous currents are not created. A circular stainless steel screen inserted over the inflow and siphon tubes prevents tadpoles from becoming trapped in the neck of the bottle. Tadpole bottles, each containing 25 to 75 tadpoles, depending on size and species, are held in racks provided with waste troughs to dispose of the water which is siphoned out.

In nature, certain individuals of each batch of tadpoles grow most rapidly at first. These large individuals release a growth-inhibiting substance which acts upon the smaller tadpoles. The result is that tadpoles metamorphose and emerge as frogs in waves, rather than all at once. While in nature this helps avert mass mortality due to predation or unfavourable conditions in the terrestrial environment, commercial culturists would not be likely to find this arrangement advantageous. Fortunately, the use of continuously flowing *water,* as described here, prevents the accumulation of the growth-inhibiting substance and results in relatively uniform growth and more or less simultaneous metamorphosis.

Cages for Metamorphosed and Adult Frogs

Metamorphosis is considered to have occurred when the forelimbs erupt, at which time the young frogs are transferred to rectangular plastic containers lined with rubber mats and containing a few pieces of broken clay flower pots to serve as cover. These cages are placed on racks at an incline, so that one end contains water while the other is dry. The containers are cleaned and the water is changed every third day.

Transformation into the frog form need not be complete for the young to be treated as adults. Rather they are transferred to adult containers as soon as their vigor is assured. Housing for adult frogs should have the following characteristics:

1. Both aquatic and terrestrial areas.
2. Flowing water and facilities to permit flushing.
3. Construction design which permits ready access and allows cleaning with minimai handling of the frogs.
4. Closures which are strong enough to prevent the escape of frogs, fine enough to retain insects presented as food, and open enough to provide adequate ventilation.

These conditions are met by the rather elaborate two-story plastic cages used at Michigan. The opaque bottom section of each cage contains not only water but ledges so that the frogs may remain dry but secluded. Nested into the bottom sector is a dry, transparent cage with a rubber mat and clay shelters like those used in containers for newly metamorphosed frogs. A hole in the bottom of the upper section permits the frogs to move from level to level. Water is supplied through a tube in the rear of the bottom compartment and leaves via a 25-mm-high overflow tube. Tops are made of stainless steel wire cloth and incorporate an access port.

Adult cages, which are either 48.3 cm x 26.7 cm x 16.5 cm deep or 50.8 cm x 40.6 cm x 21.6 cm deep, are mounted on racks so that each cage can be individually withdrawn for inspection, cleaning, and so on. The overflow tubes are connected to a telescoping drain so that circulation need not be interrupted during such operations.

A simpler device for holding adults, which may be more readily adaptable to commercial culture, is a 5-m-long trough, equipped similarly to the cages just described, but not incorporating the two-story structure. Such troughs may be divided into compartments, without losing the desirable feature of easy flushing, by simply leaving a 6-mm open space under each divider.

Separate housing facilities are maintained as a quarantine for newly received frogs. Such animals are first held 10 min in water containing enough calcium hypochlorite to produce a chlorine concentration of 6 ppm. Following this treatment, they are placed in isolated containers and provided with an excess of food until their health is assured.

Feeding

Feeding is critical to the success of all forms of frog culture, and Nace's system is no exception. Tadpoles may be fed a variety of greens, but boiled romaine or escarole lettuce has been found best. Spinach is avoided, as it may cause formation of kidney stones. A good deal of judgement is required in feeding, for, although tadpoles consume great amounts of food and must be fed twice daily, they may be killed by over feeding. The basic lettuce diet is supplemented two or three times weekly with cubes of raw or boiled liver. It should be noted that a number of other forms of food may be equally suitable from the tadpole's point of view but that the nature of the tadpole bottles dictates a number of the properties of the food. Specifically, it must not float to the surface, produce a scum which inhibits gas exchange, or be fine enough to settle through the screen into the neck of the bottle or be flushed out.

As in outdoor culture methods, feeding presents more of a challenge once metamorphosis is reached. So far, optimal growth and rapid attainment of sexual maturity have been attained only when the frogs are fed on live insects. Nace has settled on three species, the meat fly *(Sarcophaga bullata)*, the greenbottle fly *(Phoenicia sericata)*, and the field cricket *(Acheta domestica)*. Crickets are obtained commercially and maintained on a diet of chick mash and water, but the flies are raised in the laboratory.

Adult flies are maintained on water, sugar, and a sugar solution and allowed to deposit their eggs on a moistened mixture of sawdust and dog food, topped with several thin slices of raw liver, placed in a plastic tray. After 24 hours in a breeding cage, each such tray is placed in a 31.0 cm x 28.5 cm x 8.1 cm deep stainless steel pan, lined with paper toweling. Escape of maggots is prevented by means of a nonlethal "electric fence" created by running a 10-V current through a copper strip mounted on insulation affixed around the lip of the pan. Upon reaching full growth, the maggots migrate from the food tray into the pan, where they pupate. Pupae may be stored at 4°C for as long as 3 to 4 months, then warmed to 30°C for hatching. After hatching, the flies may again be chilled or anesthetized with CO_2 and fed to the frogs while in a torpid state.

The entire fly culture operation, which produces 25,000 flies daily, is confined to a 2.4 m x 3.0 m room. The species raised both require elevated temperatures for reproduction, thus escapees cannot become a nuisance in the Michigan climate. Potential frog culturists in tropical and subtropical climates, where meat flies and greenbottle flies might become established, might consider the possibility of relying entirely on crickets or some other species of insect.

The diet of flies and crickets has proved satisfactory for green frogs and leopard frogs, but pickerel frogs and some smaller species apparently develop a vitamin deficiency and do not survive well. Ultraviolet lighting has been found to help, but it is more efficient and equally effective to dust the food animals with powdered Pervinal, a commercial preparation containing ten vitamins, as well as calcium, magnesium, and various trace elements.

Breeding and Selection

Laboratory breeding procedures for frogs have been standardized and are well described in a number of embryology texts, thus need not be repeated here. Commercial adaptations have yet to be worked out.

Frogs have been selectively bred for some time, a process which has been greatly enhanced by the discovery that the skin markings of each frog are unique and may be used like human fingerprints as a means of individual identification. All frogs in the Michigan colony are routinely identified at metamorphosis and a complete breeding record kept for each individual, a procedure which could easily be applied by commercial culturists to selected breeder frogs.

Possibility of Indoor Commercial Culture

Obviously, the methods just described will have to be simplified somewhat if frogs are to be commercially cultivated indoors. The first to attempt to adapt Nace's and Kawamura's techniques to commercial culture and also the first to culture bullfrogs indoors, is Duddley D. Culley, of the Department of Forestry and Wildlife Management, Louisiana State University. The principal difference between Culley's procedure and that used at Michigan is the source of food. Tadpoles are fed on commercially available rabbit pellets or trout chow. Adult bullfrogs have been found to prefer fish to other food organisms, thus experimental animals at Louisiana State are maintained on a diet of mosquito fish *(Gambusia offinis)* and saifin mollies *(Mollienisia latipinna)*, both of which are easily bred and maintained in captivity. Bullfrogs are also fond of tadpoles, and experiments are being conducted with excess tadpoles as food. The principal obstacle to total biological success in indoor bullfrog culture is the difficulty of breeding them, but Culley is optimistic that this problem will eventually be solved.

Diseases

Diseases present a relatively small problem in frog culture. The most commonly reported "disease" is "red-leg," usually attributed to overcrowding. Actually there are two conditions which may give rise to a reddish discolouration of the legs. Red-leg in recently transported frogs most often indicates simple irritation of the skin caused by prolonged contact with a dry surface. Such irritation may, if not treated, afford access to infectious microbes, but it is not a disease symptom in itself. Infection by certain bacteria, most frequently *Aeromonas,* produces a similar response. The best preventive measure is adequate nutrition. Cures may be effected by isolating the infected individuals and treating them with such antibiotics as chloramphenicol and sulfadiazine. In severe cases, it is also advisable to keep the frogs in a salt solution approximating 25 to 30% frog Ringer's solution.

Growth and Development

Growth and development of well-fed frogs in both outdoor

and indoor culture systems compares favourably with that observed in nature. The chief determining factors are food supply and length of the growing season. On the average, two years are required from metamorphosis to maturity in the South and four years in the North. Similar variations exist in the growth rates of tadpoles.

Of perhaps more interest to the culturist is the time required from metamorphosis to marketable size. The only reliable data of this sort come from Culley's experiments. Taking a length of 20 cm (including the outstreched legs) and a weight of 130 g as minimal for marketing in Louisiana, almost all of Culley's bullfrogs reached commercial size within 12 months of metamorphosis. The fasters growing individuals reached this size in 8 months. Through selective breeding, Culley hopes to reduce the average time required to less than 8 months.

Utilization

A serious obstacle to the development of commercial frog culture in the United States is the American attitude that only the hind legs are useful as food. There is in fact an ample portion of meat on the back and front legs of any large frog, but its utilization would increase the cost of processing. Even if the back and front legs are eaten, there is still a large amount of waste. In certain Oriental cultures frogs are prepared so that the bones become soft and digestible, or even cooked with the entrails and skins intact, but it is doubtful that such practices would find favour in the United States at this time.

Culinary practices aside, if markets or uses could be found for presently wasted parts of the frog, frog culture in the United States would be closer to the threshold of economic feasibility. Mention has already been made of the practice of feeding frog entrails to tadpoles. It is also possible that frog wastes could be processed into a product similar to the marine protein concentrate (MPC) made from fish, and used as an animal feed or nutritional supplement. Efforts have also been made, with some success, to tan frog skin and use it in the manufacture of leather items.

In addition to the economic and gastronomic incentives, frog culture should be encouraged as a conservation measure. Wherever there are large populations of frogs in the United States, they are sought by hunters not only as a source of food but for sale to schools and research laboratories. Intensive hunting, along with drainage of wetlands, continues to reduce already depleted populations and some authorities foresee the disappearance of the wild frog industry within 10 years. Already American educators and scientists must import large quantities of frogs.

Disappearance or drastic depletion of frogs would mean the loss not only of an industry but of an important component of the ecosystem. Adult frogs are among the most effective insect predators, and both adults and tadpoles are important in the diets of many fish, birds, reptiles, and mammals. Tadpoles occupy a unique position in the food chain by virtue of their benthic feeding habits, which result in their recycling nutrients that might otherwise be trapped in the substrates of ponds. It is thus to be hoped that aquaculturists in the United States will make increased efforts to propagate and rear frogs for both laboratory and table use.

Experimental Frog Culture in India

Frogs support industries of some value in many of the Latin American and Asian countries, but only India maintains an extensive frog culture program, although small-scale experimental culture is reportedly being carried out in China and Cuba. The main center of experimental frog culture in India is the Pond Culture Substation of the Central Inland Fisheries Research Institute, Barrackpore, Cuttack, but preliminary work is also being done at the Freshwater Biological Station, Bhavanisagar, Madras. In both cases the species cultured are *Rana hexadactyla* and *Rana tigrina,* both of which are in high demand for export as frog legs. Officials at Cuttack are not generous with information, but it is known that some of the studies conducted concern induced spawning food and habitat requirements, and polyculture with fish.

Spawning

Both species of frog, which spawn naturally during the northeast monsoon (September to November), can be induced to spawn throughout the year by the administration of frog pituitary hormones with a priming dose of progesterone. The natural rate of fertilization of *R. tigrina* is poor, so stripping of eggs and artificial fertilization have been employed. Nearly 100% fertilization has been obtained using the dry method.

There appears to be no basis for fear that either species will be a serious predator in fish ponds. Preliminary observations indicate that they feed primarily on worms, gastropods, and aquatic insects, with small fish constituting only an incidental item in the diet. If submerged weeds are encouraged by fertilization, a frog pond should produce an ample food supply.

Polyculture with Fish

Indications are that the combination of frogs and fish would be more profitable than frog monoculture. Experimental yields obtained when small frogs were stocked with fingerlings of the Indian major carps catla *(Catla catla)*, rohu *(Labeo rohita)*, and mrigal *(Cirrhina mrigala)* are shown in Table 13.1.

Table 13.1 Stocking Rates and Yields of Fish-Frog Culture in India

Frog species stocked	*Stocking rate/ha*	*Stocking rate of major carps per hectare*	*Yield of frogs (kg/ha)*	*Yield of fish*	*Total Yield*
R. hexadactyla	2,000	none	259.0	0.0	259.0
R. tigrina	2,000	none	235.6	0.0	235.6
None	–	3,705	0.0	886.1	886.1
R. hexadactyla	2,000	3,705	234.8	1,611.0	1,815.8
R. tigrina	2,000	3,705	218.3	1,093.1	1,311.4

No explanation has been advanced for the higher yields of fish in ponds stocked with frogs, and they may or may not be related to the presence of frogs.

14

Culture of Tilapia

Members of the genus tilapia (family Cichlidae) have been an important source of food for man at least since recorded history began. The fish Saint Peter caught in the Sea Galilee and those with which Christ fed the multitudes were tilapia. An Egyptian tomb frieze, dated at 2500 B.C., illustrates the harvest of tilapia and suggests that they may have been cultured. Since that time, and probably before, the various species of tilapia have been of major importance in the fisheries of their native lands, the Near East and Africa.

Distribution of *Tilapia* Spp.

From the point of view of human nutrition, tilapia was already firmly entrenched as one of the world's most important fish by the start of the twentieth century. But with greatly increased emphasis on fish culture in this century, plus the advent of modem transportation, tilapia became even more valuable to man. Today no fish, with the probable exception of the common carp *(Cyprinus carpio)*, is more widely cultured. As early as the 1920s experiments in tilapia culture were being carried out in Kenya, Tilapia had become an intercontinental traveler by 1939 when naturally propagating stocks of *Tilapia mossambica*, native to the streams of Africa's east coast, were discovered in Java. No one is quite sure how they got there, but it is likely that aquarists, who have long been intrigued by the bizarre "mouth-brooding" habits of this and most other tilapia species, were implicated. However T. *mossambica* arrived, they made themselves at home and spread rapidly throughout the island. So prevalent did they become that, despite the geographic reference in the scientific name, the generally accepted common name for T. *mossambica* is "Java tilapia." They were originally regarded as a nuisance, but when facilities for the traditional Indonesian practice of milkfish *(Chanos chanos)* culture began to deteriorate under the Japanese occupation, the less demanding nature of tilapia became apparent, and by the end of World War II tilapia were not only too well established in the waters of Java but too deeply entrenched in Indonesian fish culture for anyone to consider control measures against them.

The Japanese, always a nation cognizant of the nutritional possibilities of fish, helped spread tilapia throughout Indonesia, where today it is found in virtually every body of water, including ditches and stagnant pits where few other fish of value can be grown. The retreating Japanese also introduced tilapia to Malaya, from whence it spread throughout southeast Asia, achieving considerable importance as a food fish in virtually every country. Today, due to the enthusiasm generated by the research of such men as H.S. Swingle at Auburn University in Alabama and C.F. Hickling at the Tropical Fish Culture Research Institute in Malaysia, plus the missionary efforts of such organizations as F AO and the Peace Corps, the Java tilapia and its congeners are cultured, at least experimentally, not only in southeast Aisa, but in Japan, Asiatic Russia, the Indian subcontinent, the Near East, virtually all of Africa, parts of Europe, the United States, and many of the Latin American countries.

Selection of Species for Culture

The Java tilapia was the first member of the genus *Tilapia* to come to the attention of large numbers of fish culturists and has remained the most widely cultured species. However, according to the late A. Yashouv of the Fish Culture Research Station at Dor, Israel, it is "the most difficult to manage" member of the genus. This is not to say that there is a "best" tilapia for culture. At least 14 species have been cultured and all share the hardiness, ease of breeding, rapid growth, and high quality of flesh which have made the Java tilapia popular. Anyone of them or some as yet untried species or hybrid might be the right fish for culture in a given situation. Too often problems in tilapia culture have resulted from hasty or uninformed selection of a species. In choosing a species the culturist should take into account the factors listed in Table 14.1 and discussed further here.

Availability

The Java tilapia is practically universally available, but some species are still available only within their home range or have been introduced elsewhere on a limited scale. This will probably become a less important consideration as time goes by and the more desirable species are distributed among fish culturists.

Food Habits

All tilapia are more or less herbivorous, but some prefer higher plants, whereas others are adapted to feed on plankton. If the food habits of a species are not known, they may be predicted by examination of the gill rakers. Numerous long, thin, closely spaced gill rakers indicate a plankton feeder; the opposite shows that the fish consumes larger particles of food. Some tilapia are relatively omnivorous and will benefit from artificial feeding with vegetable materials and in some cases even accept animal food if nothing else is available; others are obligate herbivores. Some macrophyte feeders are sufficiently voracious to function well as biological weed controls, whereas others are useless in this capacity.

Salinity Tolerance

Most tilapia are tolerant of brackish water, but some are better adapted to it than others and may thrive and even breed in sea water.

Temprament

Tilapia are less aggressive than most carnivorous cichlids, but they may attack and nip the fish of other species, an undesirable habit if a fish is to be used in polyculture. This behaviour is not necessarily species specific; there are conflicting accounts in the literature for several species. Such factors as sex, temperature, and population density are known to affect aggression and may be involved in the reaction of tilapia to other fish.

Temperature Tolerance

Tilapia are essentially tropical lowland fish, but some species and some stocks withstand cool temperatures much better than others. Where no data are available, valid inferences may often be drawn from consideration of the climate of a species native habitat.

One more word should be said in behalf of the Java tilapia and

Table 14.1 Characteristics of *Tilapia* Spp. Used in Practical and Experimental Fish Culture

Species	*Availability*	*Food Habits*	*Salinity tolerance*	*Temperament*	*Temperature tolerance*
T. andersoni	Cultured in Katanga and Zambia; seldom mentioned in literature of fish culture	Omnivorous	Unknown	Unknown	Unknown
T. aurea	Native to West Africa from Senegal to the Chad Basin and the lower Nile, also Israel and Jordan; experimentally cultured in Alabama	Unknown	Mainly a freshwater species	Aggressive	Unknown
T. galilea	Native from Jordan to east and central Africa as far west as Liberia; cultured in Israel and the Congo	Uncertain; referred to in the literature as "omnivorous" and "strictly a plankton feeder"	May vary according to the source of stock; certainly populations in such waters as the Sea of Galilee of coastal waters would be quite tolerant, but inland populations from central Africa might not be	Unknown	Unknown; probably varies with source of stock

Contd.

(Contd. Table)					
T. heudeloti (black-chinned mouth-breeder) -considered by E. Trewavas, the leading authority on *Tilapia* taxonomy, to be synonymous with *T. macrocephala*	Range; coastal west Africa from Senegal to the Congo; widely cultured experimentally	Herbivorous, but not so voracious a feeder on higher plants as some tilapia, suggesting that it may utilize phytoplankton and / or algae	Mostly found in brackish water in nature	Generally considered peaceful with its own and other species	Does well in aquaria at 20-30°C
T. hornorum (Zanzibar tilapia)	Native to Zanzibar and the east African coast opposite Zanzibar; introduced to Malaysia by C.F. Hickling in the 1950s; more recently introduced to Costa Rica	Unknown; small numbers of gill rakers suggest that it is not a plankton feeder	Very wide in Zanzibar population	Unknown	Unknown
T. leucocisia	Cultured in Uganda, where it is native, in the 1950s; since largely supplanted by *T. mossambica*, *T. nilotica* and *T. zillii*	Unknown	Unknown; may be relatively low in cultured stocks which are *of* inland origin	Unknown	Unknown

Contd.

(Contd. Table)

T. macrocephala -considered synonymous with *T. heudeloti,* which see	–	–	–	–	
T. macrochir	Native to Congo, Zambesi, Kafue, and Okavango river systerns; has been cultured, at least experimentally, in Cameroon, the Ivory Coast, Katanga, Rhodesia, Rwanda, Madagascar, and the Sudan	Phytoplankton feeder	Probably varies with source of stock	Unknown	Growth poor in cool climates
T. melanopleura (Congo tilapia)	Native from Upper Congo to South *Af*rica; commercially cultured in South Africa and experimentally cultured in several Asian, European, and American countries	Strictly an herbivore with a preference for higher plants; may be used in weed control	Unknown	Unknown	Unknown
T. mossambica (Java tilapia)	Cultured in virtually all *of* southeast Asia, the Near East, and	Mainly a plankton feeder, but consumes all kinds *of* plants	Very wide; found in brackish as well as fresh waters in	Somewhat aggressive toward other species; this may	There is a great deal *of* experimental and observational data on

Contd.

(Contd. Table)

	southern Africa; experimentally cultured in Japan, Latin America, the United States and the Soviet Union; stocking may currently be obtained almost anywhere in the world; by far the most widely cultured *Tilapia*	and artificial feeds *of* vegetable origin; in the absence *of* plant food will accept animal food	the wild, and will even breed in sea water	be related to the deleterious effect of Java tilapia on production of catla (*Catla catla*) and, in some instances, milkfish	this species; optimum temperatures suggested range from 22 to 30°C, but they can be cultured at temperatures as low as 15.5. C, below which they cease to feed; will not survive long below 12°c, and 9°C is lethal: does not grow during the cold months in such climates as Alabama and Israel, nor above 1000 m in the tropics; its hybrids with *T. nilotica* are more cold resistant.
T. nigra	Cultured, at least experimentally, in east Africa	Omnivorous	Unknown	Unknown	Unknown
T. nilotica (Nile tilapia)	Range: from Syria into east Africa through the Congo to Liberia; widely introduced outside that range; probably next to *T. mossambica* the most commonly cultured tilapia	Variously reported in the literature to be a plankton feeder, an omnivore, and to feed on higher plants to the extent that it may be used in weed control, though not as effectively as *T. melanopleura*; requires plant food when kept in aquaria	Probably quite high since it is found in brackish water and is generally considered one of the hardiest tilapia	Little studied; may be aggressive toward other species	Similar to *T. mossambica*; does well above 15.5°C, does not survive below 12°C, lethal temperatures 11°C and 42°C; hybrids with *T. mosambica* are more cold resistant

Contd.

(*Contd. Table*)

T. randalli	Once cultured in the Congo, now largely supplanted by other species	Unknown	Unknown	Unknown	Unknown
T. sparmanni	Native to east Africa south of the equator; cultured experimentally in Japan	Omnivorous	Unknown	Aggressive toward other species	Less tolerant of high or low temperatures than *T. mossambica*; does well in aquaria at 20-30°C; optimum for breeding, 23-26°C
T. volcani	Native to Lake Rudolf, Kenya; a few specimens accidentally imported to Israel in 1967-1968; other supposed *T. nilotica* from Lake Rudolf may be *T. volcani*	Unknown	Unknown	Unknown	Unknown
T. zillii (Zill's tilapia)	Native to Near East and Africa north of the equator; cultured in east Africa and, experimentally, in Madagascar and Malaysia	Strictly herbivorous, mostly on higher plants; some use in weed control	Unknown	Aggressive toward other species	Optimum 22-24°C; aquarium stock well at 20-30°C Optimum for breeding 26°C; central African stocks do not do well below 20°C; but north African stocks withstand 14-16°C for long periods of time, although growth is greatly retarded.

Nile tilapia. Both of these species may be and are cultured where great amounts of organic enrichment make it scarcely, possible for other edible fish to survive.

Spawning and Growth of the Young

In many forms of fish culture, obtaining spawn is one of the most difficult tasks. Tilapia present no such problem; indeed it is difficult to prevent them from spawning. The fact that it takes no skill whatever to spawn tilapia in ponds is one of the reasons they have been widely promoted as a fish for subsistence culture, and certainly these prolific fish have enabled many an African or Asian farmer to produce his own fish without acquiring extensive skills or technological know-how.

To spawn tilapia little more is needed than a pond, preferably one with a loose, sandy bottom, and some breeding stock. The spawning pond should be stocked with 25 to 30 females/1000 m^2 and about half again as many males. If the water is warm enough, the males will begin digging holes, perhaps 35 cm in diameter x 6 cm deep, in the pond bottom. A female deposits 75 to 250 eggs in such a nest then picks them up in her mouth. Next the male discharges sperm into the depression and this too is picked up by the female. Fertilization thus takes place inside the female's mouth, where hatching occurs within 3 to 5 days. The larvae are retained in the mouth until the yolk sac is absorbed, after which they may venture forth, but for 10 to 15 days they still return to the female's mouth when threatened. During this time the female eats seldom if at all. If the young are separated from the mother during brooding, they may be raised quite satisfactorily by themselves. Apparently the chief function of mouth-brooding is protection from predators.

The young tilapia mature at an age of 2 to 3 months, at which time they are 6 to 10 cm long. From then on they breed every 3 to 6 weeks as long as the water is warm. Whenever the water temperature approaches the lower limits of tolerance for a particular species, breeding activity is suspended. Thus the nonreproductive period ranges from about 2 months in subtropical climates to none whatever near the equator. As for other environmental factors impinging on spawning, it can be stated that tilapia will reproduce in almost any sort of water they can survive in. Java tilapia have been successfully spawned in water with a salinity of 35%.

This description of spawning applies specifically to Java

tilapia, but it generally fits all the cultured *T. heudeloti,* in which the male incubates the eggs, *T. galilea,* in which both sexes share in mouth-brooding, and *T. sparmanni* and *T. zillii,* which spawn in typical cichlid fashion, on a clean stone or other smooth object, and do not mouth-brood, although the parents do guard the eggs and young. *T. sparmanni* and *T. zillii* compensate for the lack of the protective mouth-brooding trait by producing more eggs-up to 5000 in large *T. zillii.* In nature the survivors from a few thousand eggs in the open would probably approximate the survivors of a few hundred mouth incubated eggs and young, but in the protected environment provided by the fish culturist, more eggs are likely to mean more young. Thus *T. sparmanni* and *T. zillii* are even more prone to overpopulate a pond and produce a stunted population than are other tilapia.

The Problem of Overpopulation and Methods of Control

Mouth-brooders or not, overpopulation is the greatest problem encountered in raising any species of tilapia. Stunting is bad enough in countries such as the Philippines where tilapia as small as 75 mm may be marketed, but in places like East Africa, where large fish are preferred for the table, it can spell the difference between success and failure. The seriousness of the problem may be emphasized by describing the situation in Kivu Province of the Congo. There yields to tilapia culture of 4325 kg/ha were considered normal, which sounds quite good until it is pointed out that 70% of such yields consisted of fish less than 15 cm long. Classical methods of preventing overpopulation and stunting include separating parents and young immediately upon hatching, monosex culture, and stocking predators along with the tilapia. Selective breeding for large fish has been attempted, but with no success. Apparently environmental override genetic factors in determining size.

Separation of Parents and Young

Separation of adults and young could theoretically be accomplished by netting out the adults after spawning. When the brooders are sufficiently disturbed by nets and the like they will usually spit out the fry. The captured adults could then be placed in another pond and the spawning pond used for a rearing pond. This technique is not practicable, however, since the breeding cycles of individual adults are far from synchronous, so that in a sizable population spawning occurs more or less continuously.

It is possible to remove the young. A method practiced in Indonesia makes use of a drainable spawning pond at a higher elevation than the fry pond. When the eggs hatch, the adults are disturbed so that the larvae are released and drained into the fry pond. Different sizes of fry may then be periodically cropped from the fry pond for further use in culture.

MONOSEX CULTURE

Monosex culture is much more commonly employed. Stocks of a single sex may be obtained by sexing and sorting the stock individually. The structure of the genital papillae is indicative of sex. In males there is a single urogenital opening on the tip of the papilla; in females the genital opening is separate and located on the frontal wall of the papilla, close to the apex (Fig. 14.1). This method has the disadvantage of being laborious. Moreover, even experienced workers can usually achieve only 80 to 90% accuracy. Since all it takes is one female inadvertently introduced into a pond of males being fattened for consumption to undo all the labour involved in sexing, easier and more effective methods for obtaining monosex stocks have been sought. The best results have been achieved by hybridization. Of the many interspecific and intraspecific tilapia crosses which have been attempted, at least three have produced 100% male offspring. These are:

♂ *T. macrochir* ♀ × *T. nilotica* (a difficult cross to produce)

♂ *T. mossambica* (Zanzibar stock) x ♀ *T. nilotica*

♂ *T. hornorum* × ♀ *T. mossambica.*

Better than 98% males were also obtained from the intraspecific mating.

♂ *T. mossambica* (African stock) × ♀ *T. mossambica.*

Attempts to produce monosex stocks of tilapia have generally concentrated on obtaining males since, unlike females, males continue to grow during breeding periods. However, males have one disadvantage; Whether or not there are females present, they optimistically construct spawning nests. The favoured location for nest building is at the base of the pond bank, and the nest-building activities of a great number of males may eventually undermine the bank. For this reason there has been some speculation as to the feasibility of stocking growing ponds with females. No crosses which consistently produce high percentages of females have been reported, but at the Tropical Fish Culture Research Institute in Malacca, Malaysia, one male Java tilapia of Malaysian origin turned up which, when bred to two females of the same stock, produced 88 and 97% female offspring. Unfortunately this male was lost.

As one might gather from this discussion, the possibilities of hybridization in monosex culture have by no means been fully explored. Only a few of the possible interspecific hybrids have been produced, to say nothing of crosses between different strains of the same species, which vary in sex composition of the F_1 generation as widely as interspecific crosses. Further, the F_1 generation of many tilapia crosses are fertile and might be used in further breeding experiments. Add to this the fact that the taxonomy of tilapia is by no means cut and dried, and it can be seen that the possibilities are almost unlimited, but that duplication of a desirable hybrid may be quite difficult.

Monosex culture requires a certain amount of technical supervision to be successful. The small farmer often lacks the time, expertise, or inclination to sex fish or keep spawning records. Thus monosex culture is best suited to large commercial operations, or wherever there can be considerable governmental supervision. Thus while monosex culture of tilapia has been somewhat successful in Uganda, where the government has largely taken over fish culture, in the neighbouring Congo, where fish culture is the responsibility of the individual farmer, it has failed.

Control by Predators

Where tilapia are to be raised on small, technologically unspecialized farms, a more suitable method of population control is the stocking of predaceous fish to crop the young tilapia. The use of predatory fish is most common in Africa, where stunted fish are often simply unacceptable as food. The most commonly used predators there and in southeast Asia are catfishes of the genus *Clarias*, but eels *(Anguilla japonica)*, largemouth bass *(Micropterus salmoides)*, and carnivorous cichilds such as *Serranochromis robustus* and *Hemichromis* spp. have also been used.

Cage Culture

Experiments at several Universities suggested a completely different approach to population control; culture in floating cages.

Nile tilapia stocked at 7000 to 15,000/ha in such cages exhibited growth and survival comparable to that obtained through pond culture but were apparently unable to reproduce. It was supposed that the eggs and sperm passed through the bottom of the net. More recently, however, Java tilapia have been found to have successfully reproduced in floating cages in Lake Atitlan, Guatemala.

STOCKING SYSTEMS

Monoculture

Inclusion of a predator is one form of polyculture, a practice which is becoming nearly universal in raising tilapia. Indeed, except for very primitive subsistence culture, or in waters that will not support other edible fish, monoculture of tilapia for human consumption seems scarcely defensible. Monoculture is practiced, on a small scale, in rice fields in southeast Asia. Stocking rates for this practise are 120 to 180 fingerlings/ha. Care must be taken to use plankton and algae-feeding species, since macrophages might destroy the rice crop.

Polyculture

Since the ecological niches of the various *Tilapia* spp. are imperfectly known, and since the history of tilapia culture is so short compared to that of some other polycultural systems, species combinations and stocking practices are by no means codified. Table 14.2 outlines some of the stocking systems which have been commercially or experimentally applied, along with their results, when known.

It bodes well for the future of tilapia in fish culture that in almost every case where tilapia have been added to an existing pond culture community, total production has risen with no reduction in the non-tilapia components of the harvest. One exception was found in India where tilapia of an unknown species depressed total production of a pond and apparently virtually eradicated milkfish and catla. Where tilapia are accused of exterminating other fishes, one must wonder whether some hyperaggressive species with a more typical cichlid temperament may not have been stocked. Tilapia have not been accused of eradicating other fish in the Philippines, but milkfish farmers there are ambivalent if not hostile toward the presence of Java tilapia. Part of their dislike of tilapia stems not from any direct effect of tilapia on fish production but because tilapia, in

Table 14.2 Stocking Systems Used in Tilapia Culture, and Their Results

Country	*Species stocked*	*Type of culture*	*Results*
Cameroon	*T. nilotica, Heterotis niloticus, Hemichromis, fasciatus*	Both subsistence and commercial pond culture	Good production and effective population control of tilapia
China	*T. mossambica*, striped mullet *(Mugil cephalus)*, milkfish *(Chanos chanos)*, Chinese carp association	Traditional Chinese ponds culture	Addition to tilapia, mullet, and milkfish to Chinese carp culture increases total yield in practice, even in cases where the yield to traditional stocking was as high as 7500 kg/ha
Costa Rica	*T. melanopleura* and/or *T. mossambica*	Experimental culture in private and government-operated	Average annual production, 2,784 kg/ha; maximum 3,449 kg/ha
England	*Tilapia* sp., common carp	Experimental culture in power station cooling ponds	–
India	*Tilapia* sp., milkfish, catla *(Catla catla)*	Growing in very fertile ponds	Yields of tilapia were quite good but the total yield of the pond was depressed; milkfish and catla were virtually wiped out, presumably by the tilapia
	Tilapia sp., *Barbus* sp., Nagendram fish *(Osteochilus thomassi)*	Commercial freshwater pond culture	–
Indonesia	*T. mossambica* (35%), common carp (30%), nilem *(Osteochilus hasselti)* (20%), gourami *(Osphronemus goramy)* (15%)	Short-term growing in small tropical ponds	Good production of small size fish

Contd.

(Contd. Table)

Israel	*T. nilotica* (fingerlings at 2,550-3,000/ha), common carp (2,500/ha)	Experimental pond culture	Average total yield to monoculture of carp, 4,159 kg/(ha) (yr); average total yield with tilapia added, 5,292 kg/(ha) (yr)
	T. nilotica (3,000/ha), common carp (3,500/ha)	Experimental pond culture	Yield of carp same for monoculture of polyculture, but addition of tilapia raised total yield by 13-35%
	T. nilotica, common carp	Experimental pond culture with fertilization and supplementary feeding	Monoculture of carp yielded 900-1,500 kg/ha; total production with tilapia added was over 2,500 kg/ha
	T. nilotica (1,000-1,500 fingerlings/ha), common carp (2 size groups)	Commercial pond culture	Total yield higher than can be achieved by monoculture of either species
	T. nilotica, common carp, *Mugil cephalus, Mugil capito*	Experimental pond culture	Addition of tilapia to carp-mullet ponds generally increased production of carp and mullet and invariably increased total production
Kenya, Uganda, and Tanzania	*Tilapia* sp., mirror carp (var.	Freshwater pond culture	Carp recommended by Y. Pruginin to increase yields of ponds presently devoted to monoculture of tilapia
Malaysia	*T. mossambica, Barbus gonionotus*	Experimental freshwater pond culture with 40-120 kg/ha superphosphate	Efficient utilization of very heavy phytoplankton blooms induced by fertilization
Philippines	*T. mossambica*, milkfish	Brackish water pond culture primarily for milkfish	Tilapia enter milkfish ponds accidentally and are often unwanted by milkfish culturists; nevertheless they may increase total production

Contd.

(Contd. Table)

Southeast Asia	*Tilapia* sp., common carp, *Barbus* sp.	Brackish water pond culture	—
	Tilapia sp., common carp, kissing gourami *(Helostoma temmincki)*	Sewage ponds	—
Sudan	*T. macrochir, T. melanopleura* (mixed ages)	Freshwater pond culture	Polyculture resulted in increased yields of both species, as compared to monoculture
Taiwan	*T. mossambica* (50% or more of harvest), *Mugil cephalus* (12%), silver carp *(Hypophthalmichthys molitrix)* (10%), remaining 25% made up of: common carp, goldfish *(Carassius auratus)*, big *(Aristichthys nobilis)*, eel *(Anguilla japonica)*, milkfish	Fresh or brackish water pond culture in very rich, often organically polluted ponds	Good production of fish
Uganda	*T. mossambica*, common carp	Freshwater pond culture	Yields to polyculture higher than to monoculture of either species
U.S.A.	*T. mossambica* (4,400/ha), channel catfish *(Ictalurus punctatus)* (1,250/ha)	Experimental culture with feeding of catfish	Production with monoculture of channel catfish, 1,400 kg/ha; production with polyculture, 1,568 kg/ha of catfish, plus 266 kg/ha of tilapia; feed conversion quotient of catfish same in either case
	T. nilotica (2,500/ha), channel catfish (7,500/ha)	Experimental culture with feeding of catfish	Production of catfish equal to or better than achieved by monoculture; total production better with addition of tilapia; food conversion quotient of catfish same in either case

Contd.

(Contd. Table)

	T. mossambica (2,500/ha), *T. nilotica* (2,500/ha), largemouth bass (*Micropterus salmoides*) (500/ha)	Experimental freshwater pond culture	Production, 21.2 kg/ha of bass and 2,118.0 kg/ha of tilapia
	T. mossambica (2,500/ha), *T. nilotica* (2,500/ha), *T. melanopleura* (2,500/ha), largemouth bass (500/ha)	Experimental freshwater pond culture, with addition of an unknown number of fat-head minnows (*Pimephales promelas*) as food for bass	Production, 108.6 kg/ha of bass and 2,182.7 kg/ha of tilapia; fathead minnows all consumed

addition to consuming phytoplankton and naturally occurring algae, eagerly devour the elaborately cultured "lab-lab" meant for the milkfish. In other countries, though, there are numerous examples of tilapia being deliberately stocked with milkfish, usually with good results.

Surprisingly little work has been done on the combined use of two or more *Tilapia* species. Although an all-tilapia counterpart of Chinese carp culture may never be developed, certainly the variety of food habits among tilapia-plankton feeders, filamentous algae feeders, macrophages, and omnivores-points in this direction. The association in Africa of the plankton-feeding *T. macrochir* and the macrophage *T. melanopleura* also usually involves an assortment of ages and sizes, not to gain any particular polycultural advantage but because, given the prolific nature of tilapia, it is simpler. Initial stocking of growing ponds is carried out with a mixture of fry, large breeders, and various in-between ages. At harvest time all but a few of the breeders are taken for consumption. Excess fry are also harvested for consumption or, more often, for use in stocking other ponds. Restocking may not be necessary, but if artificial feeding is employed, 10 to 20% of the estimated production by weight may be stocked after harvest.

Granted the relative simplicity of mixed-age culture, a harvest consisting of many sizes of fish may be economically disadvantageous. Stocking fish of approximately the same size will not produce a harvest of even-size fish, even if monosex culture is employed, but may more closely approach it. M. Huet of the University of Louvain, Belgium, has devised a formula for restocking single-size artificially fed tilapia ponds:

$$\text{number of tilapia to be stocked} = \frac{\text{total production}}{\text{individual growth}} + (\text{waste})\left(\frac{\text{total production}}{\text{individual growth}}\right)$$

Individual growth of fish may be fixed within certain limits by referring to the normal growth of the species cultured, or it may be determined more precisely by weighing a sample of the stock at stocking and again at harvest.

Waste percentage is largely a matter of estimation or perhaps intuition. Values normally used in the application of Huet's formula are 10, 15, and 20%.

Total production equals natural production plus production due to feeding. Natural production must be known from previous culture, without feeding, in the pond to be stocked or a similar pond. Production from feeding may be obtained by dividing the weight of food used by its nutritional quotient. Huet does not give a method for determining production due to fertilization, but many of the commonly used organic fertilizers double as feeds and may perhaps be treated on that basis.

As an example of Huet's formula in application, let us suppose a 0.5 ha growing pond which is cropped twice a year. Its natural production is estimated at 400 kg/(ha)(yr). Now 3000 kg of cottonseed oil cakes, which have a nutritive quotient of 5, are available as feed, and waste is reckoned at 15%. Total production for 6 months thus equals:

$$\frac{400\text{kg}/(\text{ha})(\text{yr}) \times 0.5 \text{ ha}}{2} + \frac{3000 \text{ kg food}}{5} = 700\text{kg}$$

Individual growth of fish is estimated at 0.1 kg and waste at 15%. So, plugging in the data, we get

$$\text{no. of fry to be stocked} = \frac{700}{0.1} + (0.15)\frac{700}{0.1} = 8050 \text{ fry}$$

The formula is of course applicable only if the pond is to be restocked with the same size fish at the same species ratio as originally stocked.

The only other noteworthy example of multiple-species tilapia stocking is H.W. Swingle's work at Auburn University in Alabama. In addition to two or three species of *Tilapia,* one of which, *T. nilotica,* has unknown feeding habits, Swingle used a presumptive predator, the largemouth bass *(Micropterus salmoides).* The bass may have gotten some nutrition from young tilapia, but when fathead minnows *(Pimephales promelas)* were added to the pond, they were eradicated by the bass over the course of 7 months, resulting in a fivefold increase in bass production but only a slight increase in tilapia production. It has long been part of fishermen's lore, though not well substantiated, that many predators, including the largemouth bass, are reluctant to take spiny finned prey. If so, then perhaps the role of some predators in tilapia culture should be re–evaluated.

Swingle has also stocked Java tilapia and Nile tilapia separately with channel catfish *(Ictalurus punctatus)*, which though not as piscivorous as the largemouth bass, might conceivably act as a predator on tilapia fry. Significant consumption of tilapia fry by the catfish was not noted, but the tilapia apparently utilized not only plankton, but wastes and excess feeds intended for the catfish, and production of catfish was substantially increased.

Stocking Rates

Most of the reported stocking densities for tilapia are lower than necessary, at least where monosex culture is employed. The problem of excess small fish in tilapia ponds is more closely allied to excess reproduction than to initial overstocking. Experiments at Auburn University in which 1 year-old, 100 g *T. nilotica* were stocked in 2.02 ha ponds at various rates demonstrated increased production at each density increment up to 5039 /ha, the highest tested.

Research in Uganda on the hybrid ♂ *T. mossambica* (Zanzibar stock) × ♀ *T. nilotica* (Lake Albert stock) showed that, up to a weight of 50 g, fry did not suffer retarded growth at densities as high as 8000/ha. Above that size it was found necessary to transfer them to rearing ponds at 1000 to 1500/ha if normal growth was to continue.

Much higher stocking densities were successfully employed in further experiments at Auburn University involving tilapia fingerlings. Nile tilapia fingerlings stocked at 20,000/ha and fed gave a production of 2822 kg/ ha in 196 days, but those stocked at 40,000/ha produced 3699 kg/ha. Of these fish, 98.5% and 91.9%, respectively, were considered to be of usable size. Similar results were obtained with Java tilapia at densities up to 50,000/ha, but the percentage of usable fish, 99.7 at 10,000 fish/ha, declined to 67.7 at 50,000/ha, due presumably to the very prolific nature of Java tilapia. Of course it should not be assumed that more is better than less in tilapia stocking. Many factors other than density enter the picture, not the least being the question of available food supplies and its corollary, the economics of feeding.

MANAGEMENT OF TILAPIA PONDS

Segregation of Age Groups

The practice of segregating different ages of fish in nursing, rearing and growing ponds, so prevalent in culture of species where spawning is strictly controlled, is rare in tilapia culture. However, in parts of Africa, fry may be nursed from 6 weeks to 2 months, or until they begin to breed, in 0.01-to 0.1-ha ponds. This technique may become more wide-spread, particularly for use in monosex culture.

Pond Fertilization

Pond fertilization is of vital importance in culture of tilapia, particularly the plankton-feeding species, but has yet to be treated systematically. This is largely due to the short history of tilapia culture, as well as to the frequent use of tilapia in "crash" programs designed not to provide data for more sophisticated efforts but to alleviate existing critical protein shortages. In southeast Asia, tilapia may be cultured in ponds which are already heavily enriched by agricultural runoff and/ or domestic pollution. Some of these waters carry such heavy loads of organic nutrients as to be uninhabitable by other desirable fishes. Fertilization would thus be superfluous, and fertilizers are set aside for use in culture of carps, milkfish, mullet, and so on. But even in southeast Asia, such ponds are the exception rather than the rule.

It may be expected that as tilapia culture matures, especially if tilapia becomes a commercial product sought by all economic classes, fertilization will be studied in some detail.

A certain amount of research on fertilization of tilapia ponds has been carried out, most of it involving Java tilapia and all of it involving the use of phosphates, which are most effective group of fertilizers for enhancing phytoplankton production. The importance of phosphorus in production of Java tilapia was demonstrated by a series of experiments at Auburn University in which ponds received 8-8-2 (N-P-K) fertilization, 0-8-2 (N-P-K) fertilization, or no fertilization at all. Both fertilizers significantly increased tilapia production at population densities of 4942, 9884, 14,826, and 19,768/ ha. Except at the highest density, the 0-8-2 (N-P-K) mixture was actually more effective than the mixture containing nitrogen compounds. Although it can be predicted that use of phosphates will enhance phytoplankton production, the digestibility of various phytoplankton differs widely. Java tilapia digest *Anabaenopsis* and *Oedogonium* well, *Botryococcus* partially, *Microcystis* and *Spirogyra* poorly, and perhaps cannot digest *Oscillatioria* or *Anabaena* at all.

Unfortunately it is not yet practical to produce cultures of a particular phytoplankton species on a large scale.

Experiments in South Africa yielded the none too surprising result that 185 kg/ha of 19% superphosphate in conjunction with lime increased yields of Java tilapia by a factor of greater than 4. Basic slag, which is often used in practical fish culture in Africa, achieved similar results when applied at 225 kg/ha. More surprising is the similar effect of experimental application of 19% superphosphate at 330 kg/ha to ponds containing the macrophagous *T. melanopleura.* Over a 5-month period an increase in production of 240 kg/ha over natural production was recorded.

Phosphatic fertilization in the form of P_2O_5 applied at 40 to 120 kg/(ha)(yr) to Malaysian ponds containing Java tilapia and *Barbus gonionotus* raised the total yield of fish by 261 to 1260 kg/ (ha)(yr). It should be pointed out, though, that both of these species are plankton feeders and highly tolerant of turbidity.

The effects of organic fertilizers on tilapia are not well known, but such fertilizers are fairly widely used, for example, in Indonesia, where marigold plant is used as a green manure. It is also likely that where such artificial feeds as oil seed cakes are used in tilapia culture, a large percentage of the intended feed is not consumed by the fish but functions as a fertilizer. Sewage is used to fertilize tilapia ponds in southeast Asia and has been suggested for use in semiarid regions of Africa where green manures are not abundant and animal manures are better employed in agriculture. This suggestion has met with little enthusiasm since unlike most Asians, Africans in general are as squeamish with regard to human wastes as are Europeans and Americans.

Pond fertilization in fish culture is of course not merely a matter of knowing the characteristics of the fish to be cultured, but must also take into account the chemistry of the water and soil. Thus fertilization experiments carried out with Java tilapia in, say, Alabama, are of only limited value to the prospective tilapia culturist in Uganda. As time goes on and more on-site research is carried out, scientists and fish culturists will be in a better position to evaluate the pros and cons of fertilization in culture of tilapia and all fishes.

Supplementary Feeding

Supplementary feeding in tilapia culture is as important and as little understood as fertilization. Most authorities are of the opinion that supplementary feeding is essential for success in large-scale culture of tilapia, but nowhere is it carried out in a systematic manner. Among the feeds employed in Asia and Africa are rice bran, broken rice, oil cakes, flour, corn meal, kitchen refuse, rotten fruit, coffee pulp, and a variety of aquatic and terrestrial plants. In South Africa, *T. melanopleura* are provided with fodder by growing rice to a height of 20 to 30 cm, then flooding it and allowing access to the fish.

In countries where little money is available to purchase feeds, use of waste products should be encouraged. For example, in the Congo the simple expedient of throwing mill sweepings into tilapia ponds has resulted in production well above normal for the area.

There is great need for research in tilapia nutrition, not only to determine what feeds are best but also what rates of feeding are most effective. The only worthwhile study of tilapia feeding rates, was carried out on Java tilapia and Nile tilapia at Auburn University, using a mixture of 35% peanut meal, 35% soybean meal, 20% ground beef liver, 15% fish meal, and 15% distiller's dry solubles. Experimental feeding rates were 1, 2, 3 and 4% of body weight per day. Java tilapia grew best at 3% but nearly as well at 2%. Growth of Nile tilapia improved with each increment of feed. Feed conversion rates were best at 2% and 1%, respectively. The feed used seems unusual for tilapia in its high animal content; certainly such a feed would not be economically feasible where tilapia are raised to offset protein deficiencies in human diets.

Use of Tilapia in Weed Control

Correlated with the feeding habits of the various *Tilapia* species is their use in weed control. Five species have to date been used or tested in this capacity.

T. heudeloti fed on higher plants in experiments at Auburn University, but they did not exert effective control over any species.

T. melanopleura vies with the grass carp as one of the best agents of biological control of aquatic weeds.

T. mossambica eats filamentous algae, a principal habitat for many species of mosquito larvae, thus is often used, in combination with insectivorous species, as an agent of malaria control. For this purpose, 2500 to 5000 fish/ha are recommended. Its role in relation to higher plants is not clear. Some authorities claim that it does not consume them in appreciable quantities. However, in ponds in Texas it appeared to effectively control some aquatic and emergent plants, though it would not touch *Elodea* or *Riccia*. Consumption of higher plants probably depends in part on the availability of other foods.

T. nilotica experimentally stocked at 2500 to 500/ha controlled filamentous algae and reduced some higher plants. It may be even more effective in malaria control than *T. mossambica* since it not only eats algae but is reputedly fond of mosquito larvae.

T. zillii in experiments in Malaysia controlled higher plants, including *Fimbristylis acuminata,* which grass carp will not touch.

Weed control by tilapia is generally a matter of stocking them in ponds where weeds interfere with fish culture. Thus stocked, they not only perform a service by eating weeds, but add to the total fish production of the pond. Tilapia have occasionally been suggested for use in control of nuisance plants in natural waterway where they are not native, as in the St. John's River of Florida where water hyacinth interferes with sport fishing and boating. Such suggestions are usually resisted, since the total effect of a voracious herbivore in a new environment is, to say the least, difficult to predict.

Growth and Production

Growth of tilapia varies greatly with stocking density, frequency of spawning, and food supply. Under very favorable conditions, individual Java tilapia may reach a weight of 850 g in 1 year; in brackish water they may reach 450 g in 8 months. But in most ponds 86 to 140 g is a more realistic weight to expect after a year if the sexes are raised together. Males grow two to three times faster than females, thus monosex culture of males produces correspondingly better growth. Monosex culture of females would also improve growth by eliminating periods of no growth associated with spawning, but no data are available on this rarely practiced technique.

Heterosis is not unknown in tilapia, but the subject has been insufficiently investigated. It has been shown that both of the reciprocal crosses of Java tilapia and Nile tilapia exhibit better growth and food conversion than either parent. The same is true of both the intraspecific crosses of Java tilapia of African and Malaysian stocks.

Since the future of tilapia culture appears to be largely bound up with polyculture, it seems superfluous to discuss production at length. The question to be asked is not "How many kilograms of tilapia can be produced in this pond?" but rather "Will tilapia add significantly to this pond's fish production?" As we have seen, in most cases the answer to the second question is yes. The amount of added production will of course vary, but even in instances such as the experimental culture of Java tilapia with channel catfish, where only 266 kg/ha of tilapia were produced, the effect of adding tilapia must be regarded as significant. Although if the total production of a fish pond were 266 kg/ha it would be a poor pond indeed, the production of tilapia in this case must be considered as supplementary to the production of the primary crop, channel catfish. The 266 kg/ha of tilapia produced, *plus* an increase in catfish production of 168 kg/ha over the 1400 kg/ ha achieved by monoculture, amounts to an increase of 434 kg/ha of fish, or a total production gain of 30.3%. Addition of tilapia to carp ponds in Israel has, on occasion, resulted in production gains of more than 165%.

Where monoculture of tilapia is practiced, the "normal" figure usually cited for natural production is 500 kg/ha in the tropics and somewhat less in moderate climates. With fertilization and/ or supplementary feeding yields of 1000 to 2500 kg/ha can be achieved. Huet suggests that 2500 kg/ha is a minimum desirable production in Africa but, at least in subsistence culture, smaller yields surely accrue some benefit to the culturists. Maximum yields of tilapia may be as high as 18,000 kg/ha, but reports of such yields must be taken with a grain of salt since they are likely to contain a preponderance of fish too small to be usable.

Harvesting and Marketing Tilapia

Harvesting of tilapia crops is usually done by seining or, where feasible, pond draining. However, experiments indicate that if only a portion of the stock is to be retained electrofishing may be more efficient and result in less injury to the fish.

In most places where tilapia is cultured it is marketed as fresh or iced fish, but it may also be sold frozen. The market price for tilapia varies greatly. In Israel, tilapia now brings a better price than the traditionally cultured common carp. In Africa the commercial value of tilapia is largely dependent on size. In southeast Asia, size is not an important factor, but the market price varies regionally. In some areas the acceptability of tilapia is lessened by its black skin, and consumption of tilapia is largely limited to people who cannot afford, say, milkfish. In other areas it is marketed as a gourmet food, which is quite appropriate, for the quality of tilapia flesh is usually very high. Tilapia too small to be marketed for human consumption need not be wasted, as they may be used as fodder for more expensive fish such as trout, as an ingredient in livestock feeds, or as bait in commercial fishing.

Experiments in Alabama indicate that tilapia could be produced in the United States at extremely low cost. Data are not available for other parts of the world, but it would appear that tilapia could compete favourably with other aquacultural and fishery products in most tropical and moderate climates.

Some authorities, however, notably S. Tal and C.F. Hickling, do not view tilapia as a good commercial proposition, at least not on a large scale. Even Tal and Hickling do not dispute tilapia's importance where there is an immediate need to feed large numbers of people or where fish can go directly from pond to pot. Whatever the commercial feasibility of tilapia culture may prove to be, it must be conceded that, thanks to subsistence culture and small-scale commercial culture, tilapia is currently one of the most important fish crops in much of the world, including most of Africa, Jamaica, southern Taiwan, and parts of Indonesia.

Diseases and Parasites

Diseases and parasites are somewhat less of a problem with tilapia than with many cultured fishes. Among the parasites found on tilapia are *Trichodina, Chilodon,* and *Saprolegnia.* The only disease mentioned in the literature is bacterial fin rot, but one might expect that, at least in marginal climates, tilapia would be subject to ichthyophthiriasis, the precondition for which is usually chilling. Tilapia also act as carriers of catarrhal enteritis.

Status and Prospectus of Tilapia Culture

The propectus for tilapia culture can best be outlined on a regional basis.

The Near East

The future of tilapia in Israel is debatable, but as long as the present demand is sustained it seems likely that tilapia will hold their own in fish culture as well as in fisheries. There are perhaps more institutions carrying on fish culture research in Israel than in any other country, so breakthroughs in tilapia culture techniques may well be made and applied in Israel.

The growth of fish culture in other Near Eastern countries is hampered by the lack of an aquacultural tradition and, in some cases, a shortage of water. Present emphasis in the region is on culture of the common carp, but one may expect tilapia to play some role in the development of fish culture.

Africa

There is perhaps more excitement over tilapia in tropical Africa than anywhere else, but results to date vary greatly, as does the prospectus. The problem of stunting, which must be reckoned with wherever tilapia are-raised, is crucial in Africa due to the reluctance of most Africans to accept small fish. The future of African tilapia culture is thus dependent on the improvement and spread of population control techniques.

Culture of tilapia will undoubtedly continue to be practiced in virtually all African countries but, as in the past, success may be expected to be greatest in those countries where government supervision and assistance are greatest. Most noteworthy in this respect are Uganda and the Malagasy Republic. Government supervision is necessary not only for economic reasons but because Africa lacks a tradition of fish culture. Thus a farmer who is perfectly willing to work long and hard on a terrestrial crop is culturally conditioned to think of fish as wild game rather than as livestock to be tended, and he may let a carefully constructed and stocked fish pond deteriorate for lack of attention.

Southeast Asia

Tilapia may be expected to continue to play an important role in fish culture, particularly in Taiwan and Indonesia. Whether significant expansion will occur depends on the results of research, on the extent to which eutrophication renders inland and brackish water unsuitable for culture of other fish, and, in some countries, on whether or not regional attitudes toward tilapia change.

United States

The consensus of workers at Auburn University is that tilapia are not presently feasible for culture in the United States, due to the premium placed on large fish, the difficulty in overwintering (even in Alabama, tilapia used in research must be brought indoors during the winter), the doubtfulness of consumer acceptance, the availability of a wide variety of native fishes for culture, and the justifiable concern of conservationists and sport fishermen over the possible ecological effects of the intentional or accidental stocking of tilapia in natural waterways.

Large-scale culture of tilapia is not expected in the United States but, despite the warnings of ecologists, occasional introductions may be made. Java tilapia and/or Nile tilapia have already been stocked locally as food or sport fish or for mosquito control in at least six of the southern states and Hawaii. There is as yet no indication of their impact on fisheries or the ecology in any of these areas.

Temperate Regions of Europe and Asia

Little future is seen for tilapia in these regions, except perhaps in industrial cooling waters. Experiments along these lines are being carried out in England, the Soviet Union, and West Germany. Some success has also been reported from tilapia culture in the south of France, but details are not available.

Latin America

In no other major area of the world is fish culture so poorly developed. Widespread protein deficiency in Latin America has in recent years prompted the various national governments and international aid agencies to investigate the potential of fish culture in the region. The most frequently suggested group of fishes for culture there are the *Tilapia* spp. The only serious objection which can be raised is that if tilapia find their way into natural waters, their effect on the local ecology could be disastrous. With this in mind studies have been undertaken in a number of countries, notably Brazil, Guatemala, and Peru, to determine the aquacultural potential of native fish species. Results to date are not encouraging, so experimental culture of tilapia is proceeding in many Latin American countries, especially Brazil and Costa Rica.

Commercial production of tilapia is already under way in Jamaica and Trinidad. In Jamaica tilapia culture is quite successful in terms of yield but shows little promise of being able to add significantly to the abundant fish supply contributed by marine fisheries. In Trinidad the situation is much the same as regards pond culture, but Trinidad and some other Caribbean islands have further resources for fish culture in the form of extensive fresh and brackish water swamps. It has been suggested that tilapia could be cultured and stocked in these swamps to supplement the take of fishermen who regularly exploit them.

Current and Future Research in Tilapia Culture

There is no form of fish culture that cannot benefit from research, but tilapia culture, as a young branch of the art involving 12 or so species grown under widely differing conditions all over the world, is in particular need of information obtained through research.

Growing Tilapia in Thermal Effluent

One aspect of tilapia research-the experimental culture of tilapia in industrial cooling waters-may result in still further geographical expansion of tilapia culture, but most tilapia research is aimed at improving production in the tropical and moderate climates where they are already successfully raised.

Population Control

Easily the most important advance which would be made would be a foolproof method of preventing overcrowding and stunting. This has usually been approached through hybridization. As mentioned earlier, there are at least three tilapia crosses that yield 100% male offspring, but one of these is difficult to produce, one involves the use of stocks from rather limited geographic areas, and one involves *T. hornorum,* which is not available to many culturists outside east Africa and Malaysia. What is needed is not only an easily obtained and duplicated all-male hybrid, but a series of them, including plankton feeders, macrophages, salinity and temperature tolerant strains. Such hybrids might also exhibit heterosis, as does the existing all-male hybrid ♂ *T. mossambica* (Zanzibar stock) x/♀ *T. nilotica* (Lake Albert stock).

Taxonomy and Genetics

To achieve the ends of hybridization it will be necessary to better understand tilapia genetics, both at the specific and sub specific levels; to understand, for example, why mating male *T. mossambica* from Zanzibar with female *T. nilotica* from Lake Albert produces 100% male off-spring, whereas mating the same sexes of the same species from other parts of the world may not. The attainment of this understanding will entail careful examination of the taxonomy of the genus *Tilapia,* which is currently quite confused.

Ecological and Ethological Studies

Concurrent with hybridization experiments should be studies to more precisely determine the ecological niche occupied by each of the cultured species. Many attempts at tilapia culture have yielded unsatisfactory results merely because the culturists did not have the necessary information to choose the right species or hybrid for the job. Once the roles of the various species are known, hybridization can proceed more intelligently and the culturist will have an even wider selection of tilapia types to choose from. One of the best means of approaching the problem of ecological niche would be to study tilapia behaviour and food habits under natural or seminatural conditions.

When the roles played by the various species and hybrids are known there will be a real basis for the foundation of tilapia polyculture, with or without nontilapia species, at a level of sophistication comparable to that of Chinese carp culture. Even now, studies aimed at finding better species combinations and stocking rates for pond communities would not be amiss.

Pond Fertilization and Supplementary Feeding

Among the less glamorous tasks facing researchers on tilapia culture is the assessment of bodies of water for culture. It is all very fine to do research on pond fertilization in Alabama or Israel, but to attempt to extrapolate from there to a pond of unknown physical and chemical characteristics in East Africa is scarcely realistic. For this and other purposes, regional research stations should be established wherever large scale tilapia culture is contemplated, and surveys should be made of all waters to be cultivated so that the culturist may proceed intelligently with specific management measures appropriate for his area.

Much more work needs to be done on pond fertilization and feeding in tilapia culture. Both laboratory research on nutritional needs and field work on the effects of different feeds and fertilizers in practical culture are needed. Not only the yields which may be achieved, but the protein content and marketability of the fish produced should be assessed.

With a sound basis in research, further development of tilapia culture can proceed apace. Development will surely include at least some practical culture in Latin America, gradual replacement of individually operated subsistence cultures with more efficient commercial or communal operations, education of would-be tilapia growers, particularly in Africa, as to the need for population control, proper feeding, and other management measures, and further mechanization wherever tilapia are cultured. More precise means of economic analysis of culture methods should also be applied.

At the present time there seem to be more questions than facts with regard to tilapia culture, but it is a safe bet that, for the foreseeable future at least, the tilapia complex will continue to be an important contributor to the world's protein supply.

15

Culture of True Eels (*Anguilla* spp.)

Among food fishes, true eels (*Anguilla* spp.) occupy a position in some countries roughly equivalent to that held among meat animals by turkey in the United States; they may be eaten anytime but are traditionally served on certain days. In Italy, eel is the traditional dish on Christmas Eve. Since eels are primarily a fishery product in Italy, the supply cannot be controlled, and prices go up tremendously on that day. In Japan, one day in July is set aside as "eel day," and great quantities are sold, but since culture is the dominant mode of production, prices are more nearly stable.

Whatever the season, and even in countries like the United States, where eel is not liked by the majority of the populace, it is definitely a luxury food. This status is due partly to the declining success of eel fisheries, and partly to the great expense of eel culture.

Natural History and Collection of Elvers

The only country with long history of eel culture is Japan, where *Anguilla japonica* has been farmed commercially for about 150 years and on a subsistence basis for centuries longer. It was not until after World War II that Taiwan became the second country where this species is cultured. Like all species of *Anguilla, A. japonica* has the rather unusual trait of being catadromous, that is, spawning occurs in the sea and the young migrate inshore to mature in freshwater. The inshore migration begins when the sea temperature first reaches 8 to 10°C in the spring, and the transparent, 5-to 7-cm long young, called "elvers," reach Japanese shores in December to April. Since

Japanese biologists have had very limited success in spawning eels, elvers collected at this time from streams and estuaries are the source of stock for culture.

The best runs of elvers in Japan occur in the east central part of the country, in Chiba, Ibaragi, and Shizuoka prefectures. Elvers are perhaps more abundant in Taiwan and when, in 1969, Taiwanese authorities relaxed the ban on export of elvers, 1 to 2 million were collected and sold to Japanese culturists, in addition to those purchased by Taiwanese eel growers. In both countries, elvers are captured in dip nets, kept in floating baskets or boxes for 4 or 5 days, then transferred to nursery ponds, where they are reared for 6 months to 2 years, or until they reach 12 to 15 cm in length, before sale to culturists. Some Japanese growers circumvent this expense by capturing eels of the same size from rivers, by the use of traps or bamboo pipes.

CULTURE IN JAPAN AND TAIWAN

Rearing Elvers

Nursing of elvers in Japan is carried out in two stages. First they are stocked at a rate of 300,000/ha in ponds 200 m^2 or less in area. After 3 months, the population is thinned and redistributed in 600 m^2 ponds. In both types of pond, they are fed daily with dried sardines or sardine meal, sometimes augmented by silkworm pupae. Young eels are transported in very little water in special baskets, 40 cm in diameter x 20 cm deep. From 35 to 70 kg of eels may be safely carried in such baskets, which are stacked one on top of another, packed with straw mats, and tied together. Ice may be placed under the top mat during hot weather.

Taiwanese nursery ponds are much more heavily stocked; figures of up to 3 million elvers/ ha are cited. Foods given include worms and minced, cooked trash fish, both presented in submerged bamboo mesh baskets.

Stocking regimes for eel production ponds in Japan and Taiwan are quite different from each other, in keeping with the different aquacultural traditions of the two countries. Eel culture in Japan is essentially a monoculture system, although common carp *(Cyprinus carpio)* or striped mullet *(Mugil cephalus)* may be stocked to clean up the excess food which is a necessary evil in eel ponds. In Taiwan, however, eels are integral parts of complex polyculture systems derived from the classical method of Chinese carp culture.

Stocking Production Ponds

In coastal Taiwanese ponds where Java tilapia *(Tilapia mossambica)* are the major crop, eels may constitute a minor component of the stock, but in another class of ponds they are the principal species stocked. Such ponds, which have smooth mud bottoms sloping toward the lower end, and walls lined with cement, brick, or stone, vary in size from 0.08 to 4 ha or more, but seldom average over 1.0 to 1.5 m deep. One such 4-ha pond was stocked in 1967 with-in addition to elvers at about 25,000/ha-silver carp *(Hypophthalmichthys molitrix)*, big head *(Aristichthys nobilis)*, common carp, mud carp *(Cirrhinus molitorella)*, and striped mullet. Table 15.1 illustrates the roles of each of these species in the pond ecosystem and in the harvest.

The system outlined in Table 15.1 differs from the ordinary pond polyculture systems originally developed in China and currently practiced in Taiwan and elsewhere in that it is neither the natural character of the pond nor fertilization *per se* that determines the structure of the fish community. Rather, the voracious but sloppy feeding habits of the eels dictate that large amounts of food be given, and the species and numbers of other fish stocked are in turn determined by the effect of the "waste" eel food. Striped mullet, which would not ordinarily be stocked in a freshwater pond, are of particular importance, not so much for their contribution to fish production as to control the dense growths of blue-green algae which result from the superabundance of nitrogenous matter. Other species that may be fitted into this type of ecosystem are goldfish *(Carassius auratus)*, which may replace silver carp, and crucian carp *(Carassius carassius)*, which, as omnivores, may take advantage of any niche that is not completely filled. Grass carp *(Gtenopharyngodon idellus)* may also be stocked, but the culturist will then have two species to feed, since few ponds support enough higher plants to satisfy this voracious species.

There is a trend in Taiwan toward higher stocking rates of eels. A large pond such as the one just described would probably now be stocked with double the number of elvers (50,000/ ha). Ponds less than 0.5 ha in area are stocked at higher rates-up to 160,000 elvers/ ha. The stocking rates thus far cited assume that running water can be provided. In small ponds, artificial aeration by means of pumps or mechanical paddles may also be resorted to. However, some eel ponds in Taiwan are stagnant, and oxygen deficiencies may develop. Stocking rates of such ponds must be less than half those cited.

Table 15.1 Stocking Rate and Yields of Polyculture Ponds in Taiwan, with Eels as the Primary Crop

No. stocked	*Stocked (G)*	*Size Survival (%)*	*Harvest (kg)*	*Species*	*Niche*
Eel	Principal crop, heavily fed	100,000	25	70	14,000
Silver carp	Phytoplankton feeder	8,000	10-20	50	6,000
Big head	Zooplankton feeder	1,000	10-20	100	5,600
Common carp	Utilizes excess eel food	32,000	10-20	25	4,000
Mud carp	Benthos feeder	4,000	-	100	1,600
Striped mulletgreen	Feeds on blue- algae	6,000	0.5	50	800
Total		151,000			32,000 (8,000 kg/ha)

Shizuoka Prefecture produces about 65% of the eels grown for market in Japan. Young eels purchased from nurseries are stocked in two types of water, ponds and running water enclosures. In addition, a closed recirculating system of the type developed by A. Saeki of Tokyo University, has been tested under experimental conditions but has yet to be adopted in practical eel culture.

Nearly all enclosures of both types have walls of concrete, stone, or occasionally wood. They vary greatly in size, from less than 0.3 ha to more than 3 ha, but ponds average much larger. In 1964, 84% of running water enclosures were less than 0.8 ha in area, while almost half of the eel ponds were 1.5 ha or larger. Running water enclosures are much more productive than ponds, and operating costs only about 2/3 as much, but pond operators are able to make up the difference by maintaining larger culture units.

The rate of stocking is determined by the rate of circulation of water and varies from 20,000 to 44,000 eels/ha. When common carp are added, they are stocked as 7-cm young at 10,000/ha.

Feeding

In any eel culture system, it is primarily feeding which determines the success of the operation. The more animal protein that can be supplied (at least 50% is essential), the greater the weight of eels that can be produced. Food thus accounts for an average of 51 to 55% of the expenses of growing eels for market in Japan, and may run as high as 80%.

The traditional foods for eels in Japan are fresh and cooked trash fish, silkworm pupae, and, to a lesser extent, earthworms, aquatic worms, and crushed mollusks. In Taiwan, trash fish are the most commonly used food, but offal from slaughterhouses and fish packing plants, ox blood, earthworms, aquatic worms, and small crabs, all chopped into small pieces, are also used. Crabs are believed to be especially effective in promoting growth. Taiwanese eel farmers feed silkworm pupae only in the last stages of culture, when the fish are being fattened for market. For this purpose, the pupae are soaked in water for an hour or so before feeding.

Table 15.2 Conversion Ratios of Eel Feeds Used

Country	*Food*	*Conversion ratio*
Japan	Chopped fish, silkworm pupae, etc.	5,5: 1
	Sardines	5.43-7.16 : 1
	Artificial feed (pellets)	3.03-4.36 : 1
Taiwan	Trash fish, etc.	10-15: 1
	Artificial feed (granular)	2.1-2.6 : 1
	Artificial feed mixed with fish liver oil	1.9 : 1

Conversion ratios obtained with traditional foods are extremely poor but recently developed artificial feeds have produced dramatic improvement. In Japan, artificial eel feed, which is produced in the form of pellets, contains about 60% fish meal and 20% starch, along with vitamins, minerals, and so forth. It is expensive and while it is used exclusively by the larger culturists, small operators vary their feeding plan according to the relative cost of artificial feed and fresh fish.

The artificial eel feed used in Taiwan is granular in form and is of slightly more recent origin, but has already been adopted by considerable numbers of culturists, who report it to be cheaper, easier to handle, and more sanitary than traditional foods.

Table 15.3 Composition and Chemical Analysis of the Artificial Eel Feed Used

Composition	
Fishmeal	65.00%
Defatted soybean meal	10.16
Yeast powder	10.12
Fish concentrate	5.00
Starch	8.12
Multivitamins	1.00
Lysine	0.20
Antioxidant	0.20
Binding substance	0.20
Chemical Analysis	
Crude protein	51.91%
Crude fat	5.36
Crude carbohydrate	16.03
Ash	17.97
Moisture	8.73

Whatever is fed, it is usually presented in baskets or perforated troughs, rather than being broadcast in the pond. The rate of feeding should be 5 to 10% of the body weight of eels daily, depending on temperature. Below 8 to 10°C, feeding ceases.

Growth, Yield, Production and Marketing

Even under optimum conditions of heavy feeding, high dissolved oxygen concentration, and temperatures of 20 to 28°C, eels characteristically exhibit great differences in growth between individuals (in an experiment with the Atlantic species, *Anguilla anguilla,* some individuals doubled their weight, while others, stocked at the same size, grew by a factor of 120) and special cropping systems have evolved accordingly. In Japan, about 30% of the eels stocked reach the marketable size of 100 to 200 g within 1 year after stocking. These are then harvested and replaced with an equal number of young eels. At the end of the second year, all the eels are harvested. Total survival over the 2-year period is 60 to 90%. In Taiwan, larger eels, weighing 200 g or more, are preferred by consumers. Eel ponds are selectively fished for eels on a daily to monthly basis, depending on demand, and all such specimens marketed. The other species stocked in eel ponds are harvested annually.

Yields and production of eel culture in Japan have increased considerably in the last decade. In the late 1950s, 15,000 kg/ha was considered an exceptional yield. In 1965, however, the most productive running water enclosures yielded about 45,000 kg/ha, and the average yield in running water was higher than the maximum formerly achieved. Although the lower yields realized from ponds lower the overall average yield considerably, it is still impressive.

Table 15.4 Production and Yield of Eels Cultured in Running Water Enclosures and Ponds

Type of enclosure	*Total area (ha) (kg)*	*Production (kg/ha)*	*Yield*
Running water	49	1,346,117	26,360
Pond	123	664,843	6,120
Total	172	1,810,960	10,528

About 98% of the eels produced in Japan are sold domestically, mostly as fresh fish, but also canned or as a frozen, precooked item. The remainder of the crop is exported, mostly to western Europe. There is a great demand for eels in Europe, which increases as the fisheries there decline, but Japan has not made a major effort to enter the market. To do so effectively, Japanese culturists would have to grow eels to twice the length of the 45-to 60-cm specimens currently marketed.

If the growth of eel culture in Japan has been fast, its growth in Taiwan has been spectacular. In 1965 122,753 kg were produced; by 1969 about 2 million kg were being produced annually from about 200 ha of water, for an average yield of 10,000 kg/ha. Taiwanese culturists have, since the beginning of eel culture in the country, grown somewhat larger eels than their Japanese counterparts, thus they export a significant portion of their crop.

Culture in Europe

Imports are not adequate to satisfy the high demand for eels which exists in virtually every European country and, as natural populations decline due to increased pollution of rivers, interest in culture of the native *Anguilla anguilla* grows not only in Europe but in Israel and the United Arab Republic as well. Large numbers of eels are still taken in the Italian valli and other Mediterranean lagoons, in practices some of which constitute low-intensity aquaculture, but intensive eel culture is a new phenomenon outside of Asia.

Commercial culture is already a reality in the Soviet Union and West Germany, where 20-cm elvers are collected along the coast for stocking. Culture proceeds much as in Japan or Taiwan, but *A. anguilla* is thus far reluctant to accept dry food, so trash fish are fed almost exclusively. Conversion ratios of 8 to 10 : 1 are realized, but ratios might be improved through monosex culture, since females of *A. anguilla* grow much faster than males. West German researchers have found that the shape of the culture enclosure also has an effect on growth. Eels, which are olfactory feeders, can locate food more efficiently in long, narrow ponds.

Diseases represent more of a problem with *A. anguilla* than they do with *A. japonica*. West German biologists refer to a "cauliflower" disease and a "red" disease, both of which severely affect growth and, to a lesser degree, survival, but no cures are suggested. An apparently more serious problem occurred in 1957 in the Po River delta of northern Italy, where eels constitute 70% of the production of valle fish culture. Eels there developed heavy infestation of an apparently new species of *Argulus*, and production declined by 50%. Attempts to eliminate the parasite by flushing the valli with large volumes of freshwater were only partly successful.

16

Commercial Culture of Freshwater Salmonids

The most widely cultured salmonid for sport fishery purposes is the rainbow trout (Salmo *gairdneri).* Native to the Pacific Coast drainages of North America from Alaska to Baja California, the rainbow trout has been introduced as a sport fish to virtually all suitable waters in the affluent countries. Nearly as much effort has been devoted to the propagation of the European brown trout *(Salmo trutta),* which has been distributed throughout the world and has assumed special importance in the United States, where it has been the saviour of trout fishing in many streams which have become too warm and/ or polluted to support native salmonids. The only other salmonid species which has received comparable attention from culturists is the brook trout *(Salvelinus fontinalis).* Originally restricted to northeastern North America, the brook trout has been introduced wherever water temperatures are cold enough to meet its demands, which are somewhat more restrictive than those of rainbow or brown trout.

Other species cultured for sport fishery purposes on a more limited basis include the Atlantic salmon *(Salmo salar),* of the European, coast, Iceland, and the Atlantic coast of North America from Maine north; the lake trout *(Salvelinus namaycush)* of northern North America; the cut throat trout *(Salmo clarki)* of western North America; the Sunapee trout *(Salvelinus aureolus)* of northern New England; and the most beautiful of all salmonids, the golden trout *(Salmo aguabonita)* of California's High Sierras.

Among the qualities that have rendered trout to sport fishermen is the high quality of the flesh. Thus it is not surprising to find that attempts at commercial trout culture were made as early as 1853 in the United States and perhaps earlier in Europe. Until recently, most of the techniques used in commercial trout culture were identical to those practiced in sport fish hatcheries. Though understanding of the rather different goals of the two types of trout culture has resulted in considerable differences in culture practices today, the commercial trout culturist should keep abreast of developments in the sport fish hatcheries, where many relevant new development occur. There is an extensive literature on trout culture for sport fishing, but this book is concerned only with commercial trout culture.

The "big three" species in commercial trout culture are the same three favoured by sport fishermen, but the pre-eminence of the rainbow trout is more solidly established. Of all the salmonids, none is so amenable to captivity or so tolerant of different temperatures, salinities, and population densities. The brown trout is nearly as adaptable as the rainbow trout with respect to environmental parameters but is more territorial, thus does not do as well when cultured at high densities. Some also consider it slightly less desirable as food. The brook trout grows less rapidly and is the most delicate of the there species, particularly with respect to temperature. However, it is preferred by some culturists where trout are sold by weight rather than length since it is a deeper bodied fish than rainbow or brown trout. Many fishermen also consider it a superior table fish.

Virtually all large-scale commercial trout culture is based on these three species, but in regions where salmonids enter commercial fisheries young fish of other species may be cultured and released to supplement natural reproduction for commercial as well as sport purposes.

HATCHERY TECHNIQUES

General Considerations

Wherever rainbow, brown, or brook trout are raised the basic techniques of spawning and hatching are the same. In nature, most salmonids dig nests, called redds, in streambeds in riffle areas. The eggs are deposited in these depressions, fertilized, and buried, with gravel, to hatch some months later. In no instance are cultured trout allowed to spawn in this manner. Rather, eggs are artificially obtained, fertilized, and hatched, resulting in fertilization rates of nearly 100% and hatching rates well in excess of those achieved in nature.

In general, wild rainbow trout spawn in the spring whereas brown trout and brook trout breed mainly in the fall. However, strains of these species, particularly the rainbow trout, may be found spawning at practically any time of year. Some commercial culturists take advantage of this variability by importing eggs from various parts of the world to maintain a constant supply of trout of all sizes. Others have selectively bred early or late spawning strains. Spawning time may also be regulated by artificial control of the photoperiod to simulate seasonal change or by injection of pituitary hormones. The latter technique has not found as much application among trout culturists as among growers of catfish, carp, and some other species.

The rate of hatching of eggs appears to be partially genetically controlled and brood stock is carefully selected for hatching as well as for production of large eggs. Large eggs, which are associated with the size of the female as well as the genetic strain, produce large larvae, which generally exhibit better survival than smaller ones.

Even trout of strains do not all mature at the same time and during spawning season brood females must be sorted frequently so that the very ripest individuals, as determined by how readily eggs ooze from the fish when handled, are spawned. Overripeness as well as under ripeness is a problem, for trout will seldom release eggs naturally under hatchery conditions. Use of overripe or underripe spawners may result in lower rates of fertilization and injury to the female.

Not only must brood females be handled very carefully during sorting, precautions must also be taken with their diet. Commercial feeds containing cottonseed meal adversely affect egg production and are to be strictly avoided.

Artificial Fertilization

The process of obtaining eggs and milt is called stripping and requires a fair amount of skill to avoid injuring either fish or eggs. In some hatcheries injury to the spawners is reduced by anesthetizing them with MS-222. Where anesthesia is not employed, stripping may be carried out as a one-man operation, but less fumbling and consequent injury to the fish occurs if it is done by a two-man team. One man grasps a ripe female by the caudal peduncle and pectoral fins and holds it over a pan, while the other gently extrudes the eggs by applying pressure to the abdomen with thumb and forefinger, beginning just forward of the vent and proceeding forward to the pelvic fins. If the fish is held tail down the eggs will flow naturally into the pan. Pressure should not be applied forward of the pelvic fins, since this may damage the internal organs. Another advisable precaution is to wear wool gloves so as to be able to grip the fish firmly. Care should also be taken not to break any eggs, as the albumen from broken eggs will coat other eggs and inhibit fertilization. For this reason it is not wise to try to extract every egg from a female. Forcing out the last hundred or so eggs increases the chance of eggs being broken.

As soon as a female has stripped, a ripe male is stripped into the pan using the same technique. The eggs and sperm are then gently mixed to effect fertilization by what is referred to as the "dry" method. At one time, a "wet" method was also popular, in which a small amount of water was added just prior to mixing the eggs and sperm, but the dry method is now almost universally preferred. If no eggs are broken it is customary to strip two or more pairs before emptying the pan, to eliminate the chance of fertilization not taking place due to a sterile male.

Incubation of Eggs

Hatching is best carried out in gently flowing water with a temperature of 8 to 13°C, containing at least 7 ppm of dissolved oxygen. Several devices for artificially incubating salmonid eggs have been developed, but most commercial hatcheries use a tray incubator. This consists of a series of metal or fiberglass tray stacked rather like the drawers in a dresser, so that any individual tray may be totally or partially removed. Eggs are placed in trays, a single layer to a tray, as soon as they are water hardened – within a few minutes after fertilization. Egg trays are made of a special type of wire screen with oblong openings about 15 mm x 3.5 mm. The oblong mesh retains the eggs but permits the elongate alevins to pass through.

Formerly it was necessary to inspect eggs daily and individually remove each dead one, or else fungus would gain a foothold on dead eggs and spread to smother adjacent live ones.

This tedious task has been largely eliminated by using malachite green to control fungus. The first malachite green treatment takes place before the loaded egg trays are placed in the incubator. The loaded trays are stacked four high in a trough with the bottom tray supported 25 mm. off the floor of the trough. Water flow in the trough is maintained at 24 liters/min. Each strack of trays is separated from adjoining trays by a divider. The dividers act as baffles to produce an upwelling of water through each tray. The trays are left in the trough for about 9 hours, at the beginning and end of which they are flushed with 7.6 g of a solution of 43 g of malachite green in 4 litres of water, introduced at the head of the trough. At the conclusion of the second such treatment the trays are placed in the inclubator. During this and all subsequent operations until hatching the eggs must be covered, since exposure to light may result in premature hatching or death.

Two types of incubator are used. In the drip incubator egg trays are alternated with perforated metal trays. Water is introduced at the top and allowed to drip over the eggs. This provides enough moisture to maintain the embryos but, since the trays do not retain water, the eggs must be removed to troughs just before hatching.

In the vertical flow incubator water also enters at the top but is piped from tray to tray in such a manner as to enter each tray at the bottom. Upwelling through the eggs provides adequate aeration. Since the trays containing the egg baskets are not perforated, the eggs may be allowed to hatch and alevins left in the incubator until they begin to feed.

Both types of incubator operate on a rather small volume of water, thus it is economical to control temperature to accelerate or retard development. To provide for 20 35-cm square trays, or about 600,000 to 1,350,000 eggs, 14 litres/min, is considered adequate. (The precise number of eggs depends on size and may be outside this range for exceptionally large or small eggs or for species other than those usually cultured commercially). During incubation treatment with malachite green is repeated twice weekly, using 3.75 g dissolved in 3 litres of water. Water flow is reduced to 8 litres/min during treatment and the solution is allowed to drip in at about 30 ml/ min for 1 hour.

The rate of hatching is determined by temperature. Table 16.1 shows hatching rates for the three commercially important species.

Table 16.1 Rate of Hatching, at Different Temperatures, of Eggs or Rainbow, Brown, and Brook Trout

Species	*Temperature (°C)*	*Days to hatching*
Rainbow trout	4.5	80
	7.8	48
	10.0	31
	12.0	24
	15.7	19
Brown trout	1.7	156
	4.5	100
	7.3	64
	10.0	41
Brook trout	1.7	144
	4.5	103
	7.3	68
	10.0	44
	12.0	35

Sale and Shipping of Eggs

Some culturists grow their own stock through all phases of the life cycle, but not a few hatcheries specialize in egg production since, perhaps due to the presence of trace elements or other solutes, certain localities seem better suited to producing high percentages of viable eggs. Markets for eggs include not only commercial and experimental culturists around the world but also state fish and game departments in the United States, some of which are not able to produce eggs as cheaply as commercial culturists.

Salmonid eggs are extremely tender during some periods in their development, but they are quite tolerant of handling during the first 48 hours after water hardening and again after they are eyed. Eggs are referred to as "eyed" as soon as the eye of the embryo is clearly visible-about 2 to 3 weeks after fertilization at normal temperatures. Except for extremely short journeys, eyed eggs are chosen for shipping. Eggs which have only recently become eyed are preferred since they are less likely to hatch prematurely under the relatively high temperatures experienced in even the best shipping containers.

Salmonid eggs are shipped in trays similar to those used in hatching. Since malachite green treatment can scarcely be applied en route, all dead eggs must be removed to prevent the spread of fungus. The first step in this process is "shocking," which amounts to nothing more than agitating the eggs vigorously enough to rupture the yolk membrane of any infertile eggs but not severely enough to injure viable eggs. After shocking, dead eggs turn white and can be individually removed by siphoning. The trays are then stacked one on top of another in a plastic bag with one or two trays of some absorbent material on the bottom and a tray of ice on top. Eggs must be kept moist during shipment, but water is not added because, without aeration, it would result in smothering the eggs. When so packed and placed in some sort of rigid container with good insulating properties, eggs may be shipped anywhere in the world.

Alevins are held in egg trays or in concrete or aluminum troughs until the yolk sac is absorbed. Water flow is maintained at about the level used for hatching. Once the fry become free swimming the water supply becomes crucial to the success of the operation. This is the principal reason that large-scale trout farming is confined to a few areas; most regions of the world cannot offer water in the quantity and quality needed for commercial trout culture. The minimum acceptable flow is about 3600 litres/min and considerably more is preferable. Ideally, the water should be isothermal between 10 and 18°C, have a *pH* of 7.0 to 8.5, contain at least 50 ppm of dissolved solids and 5 ppm of dissolved oxygen, and be free from pollutants. Surface water may meet most of these requirements but is almost never isothermal. Growth in even the best of trout streams is retarded by warm water in summer and/or cold water in winter.

COMMERCIAL TROUT CULTURE IN THE UNITED STATES

Water Supply

Illustrative of the potential of trout culture under optimum conditions is the world's largest trout farm, the Snake River Trout Company, located near Buhl, in southern Idaho. The owner of the farm, Bob Erkins, has tapped natural springs which supply him with 240,000 litres/min of isothermal 15°C water. Erkins believes a large part of his success is due to this water supply. Certainly it has played a role in making his property "the richest food-producing acre in the world." There is more to it, however, for the same water source has been and is still used by other culturists who do not approach Erkins' production of 600,000 kg/year with 10 acres of land. This amounts to 12% of the total United States production of trout. The following comments on commercial trout culture in the United States are drawn largely from Erkins' experience.

Stocking Density

One of the first departures from tradition made by Erkins was in stocking density. It had formerly been generally believed that in order to attain good growth trout had to have considerable room. Erkins, however, recognized that since trout culture is carried out in flowing water, stocking rates should be dependent not on the volume of water in an enclosure at a given moment but on the volume flowing through that enclosure in a given time, which is a function of current velocity.

At the present time, determination of optimum current velocity is as much a matter of art as science. If the current is too swift, energy which might go into growth will be used up in swimming. On the other hand, slack water results in the accumulation of wastes. As a rule of thumb, current should be sufficient to provide at least one complete exchange of water per hour.

The amount of trout which may be held in a given enclosure depends on many factors in addition to volume of water flow so stocking rates must be empirically determined by the individual culturists. A well managed farm should be able to sustain 100,000 kg/ha or fingerlings. Erkins gauges his operations to produce about 2.2 kg of marketable size trout per liter of water per minute. Beginning culturists should err on the low side and aim for 1.5 to 1.8 kg/litre) (min).

Since the holding capacity of an enclosure is determined by the weight rather than the number of fish in it, it is necessary to periodically thin out stock if growth is to be maintained. Concurrent with thinning the stock is size grading to equalize competition among the fish and produce more uniform fish at harvest time. Grading is done by netting the fish and placing them in a box with a slatted bottom. The smaller fish pass between the slats while the larger ones are retained. With small fish the process may be accelerated by lifting the grader out of the water and shaking it.

Some sport fish hatcheries use more sophisticated graders. The Murray-Hume automatic grader has the advantage of not necessitating removal of the fish from the water. It functions by forcing the fish to swim upstream until they meet a series of bars set at a predetermined distance apart. The smaller fish pass through, while the larger ones remain behind. Other grading devices which sort fish, into more than two size groups consist essentially of nothing more than a series of conventional graders.

Pond and Raceway Construction

Raceways 25 to 35 m long, 3 to 10 m wide, and 0.7 to 1.0 m deep are preferred to wider ponds of other shapes for rearing fingerlings. Not only is it easier to maintain a satisfactory flow of water in raceways, but eddying is reduced. Lack of eddies allows continuous flushing of metabolic wastes, which may not only cause oxygen depletion and increase the danger of disease but are claimed to be responsible for the "hatchery taste" of some cultured trout. Where it is possible to provide rapid enough exchange of water, larger ponds (up to 3 acre feet) may be used as "finishing ponds" to grow 15-cm fingerlings to marketable size.

Raceways and ponds should be so constructed that no pond flows into another pond and any pond can be drained individually. Not only is this more convenient for the culturist, it prevents the passage of metabolic wastes and disease organisms from pond to pond. Another precaution against disease which is necessary in some localities is to construct raceways and ponds of concrete. In areas where disease is not a major problem, earthern ponds have their proponents, who claim that trout derive substantial benefit from natural food organisms produced on an earthen bottom.

Before constructing any sort of trout culture facility, it must be determined whether the site is suitable. The soil should retain water *or* be suitable for concrete construction. Slope of the land should be 1 to 3% to permit an adequate flow of water and to allow for a vertical drop of at least a few inches at the inlet of each enclosure used to hold trout. The latter precaution is needed to insure adequate oxygenation at all times.

Table 16.2 Guidelines for composition of trout feed.

Component	*Amount*
Total protein	35-40%
Carbohydrate (allowable)	30%
Fat	8-10%
Fiber	4%
Vitamins per ton of feed	
Vitamin A	3,000,000 USP
Vitamin D_3	640,000 IC
Vitamin E	216,000 IU
Riboflavin	1,00,000 mg
D-calcium pantothenate	48,000 mg
Folic acid	80,000 mg
Niacin	500,000 mg
Choline chloride	1,000,000 mg
Vitamin B_{12}	20 mg
D-biotin	400 mg
Ascorbic acid	240 mg
Thiamine hydrochloride	60,000 mg
Pyridoxine hydrochloride	20,000 mg

Feeding

Early trout culturists usually fed beef liver or ground low-grade beef, and these foods may still be in use in some places. Fish offal has also been used but has often been associated with disease problems. Nevertheless, one of the most successful American trout farms, Trout Lodge Springs Hatchery, near Ephrata, Washington, feeds mainly ground carp and salmon cannery waste and has not experienced excessive disease problems. Diseases or no, unless the culturist is able to take advantage of a particularly cheap source of meat or fish, prepared dry foods pay for themselves in convenience.

Erkins was one of the first commercial trout culturists to convert to and exclusive diet of dry feed. The feed he adopted in 1953 contained wheat, fish meal, whey, cottonseed meal, and yeast. Since then a number of feed companies have come out with trout pellets containing as many as 30 to 40 ingredients, including vitamin B_{12}, antibiotics, and even paprika, to bring out the red spots of brown trout and brook trout. Most American trout farmers have followed Erkins' lead, thus the feed industry has vigorously attempted to improve trout feed formulas. This work continues even though existing feeds have produced conversion ratios as low as 1.5 : 1 in commercial culture and 1.13 : 1 in experimental culture of fingerlings. Table 16.2 lists guidelines for trout feed as determined at the U.S. Fish Farming Experimental Station in Stuttgar', Arkansas.

In practice, great variation is to be found in the protein content of trout feeds. In areas where there is a cheap and abundant supply of fish or meat, trout diets may contain as much as 60% protein. On the other hand, in the United States the tendency is to try to maximize profits by getting along with as little protein as possible. Some authorities have suggested protein levels as low as 20%.

Just what is the minimum amount of protein required to maintain trout growth and energy is not known. Research at the U.S. Bureau of Spot Fisheries and Wildlife's Eastern Fish Nutrition Laboratory at Cortland, New York, indicates that substitution of carbohydrates for protein as an energy source, which has been effective with some species of fish, does not work with trout, but that, under some circumstances, small amounts of fat may demonstrate a protein-sparing action.

Another Snake River Valley culturist, Thorleif Rangen, of Rangen, Inc., is approaching the problem of economical trout nutrition by attempting to develop a high-protein diet which contains little or no animal matter yet produces conversion ratios near 1 : 1. Thus far, 100% vegetable foods have been found lacking in essential amino acids, but chemical experiments with amino acid synthesis and the success of horticulturists in selectively breeding corn with ten times the usual amount of lysine and tryptophane suggest that Rangen's goal may be attainable.

Whereas the same feed formula may be used for trout of all sizes, particle size and rate of feeding vary according to the size of fish. Table 16.3 shows the size of food particles (commercial size designations) recommended for various sizes of trout.

Table 16.3 Particle size of dry food recommended for feeding trout of different sizes.

Size of pellet (Standard commercial Designations)	*Size of fish*
1	4,224-2,816/kg
1 and 2 mixed	2,992-2,464/kg
2	2,840-2,112/kg
2 and 3 mixed	1,936-1,056/kg
3	1,232-704/kg
3 and 4 mixed	880-352/kg
4	528-352/kg
4 and "crumbles" mixed	528-176/kg
Crumbles	352-106/kg
Crumbles and 3/32-in pellets mixed	176-70/kg
.24 cm pellets	70-20/kg
.40 cm pellets	20/kg and larger

In feeding very small trout it may be necessary to grind prepared food to producetiny particles. Grinding entails both a crushing and a cutting action. Crushing breaks down cell walls and makes the food more susceptible to leaching by the water, thus should be minimized in favour of cutting.

Suggested daily feeding portions are shown in Table 16.4. These amounts should be divided into three or four daily feedings or, for young fry, as many as ten feedings. Young fry should be fed as much as they will consume at one time or slightly more. The opposite holds true for larger fish. It is general practice in commercial trout culture to gauge feeding by the appetites of the smaller fish in a group so as to equalize the growth rate and create a more uniform product. Feeding in excess may result in slightly improved growth, but only with the loss of uniformity, and is unlikely to compensate for the increased expenditure.

A more sophisticated feeding guide, suitable for young of any salmonid species, and taking into account differences in the nutritional value of diets, was developed by David C. Haskell of the New York State Conservation Department. Table 16.5 is a slightly modified form of Haskell's guide. Before using it past records of trout growth and water temperature must be consulted to compute a "hatchery constant" as follows:

Instruction for Use of Feeding Guide

For hatcheries with constant water temperatures:

(i) A. $$\frac{L_2 - L_1}{\text{days in period}} = L$$

where: L_1= length of fish at beginning of period,

L_2= length of fish at end of period,

L= daily increase in length.

B. Hatchery constant = 300 × conversion x L

Note: Choice of a conversion should be based on recommendation of the feed manufacture or past experience with the diet.

C. Enter the feeding guide under size of fish opposite the appropriate constant and read the percentage of body weight to feed daily. *Note:* It is necessary to calculate the hatchery constant only once for each species or strain, as long as there is no change in diet or water temperature.

(ii) For hatcheries with variable water temperatures:

A. Determine average monthly water temperature in of.

B. Determine temperature units for the month. (*TU* = average monthly water temperature minus 38.6)

C. Determine the length increase per month.

D. $$\frac{\text{TU/month}}{\text{length increase/month}} = \text{TU/in.}$$

Repeat this procedure for a number of months to obtain an average *TU* per inch of growth. This figure should remain constant for any particular species or strain of trout, within the temperature range of 38.6° to 60°F, as long as the diet is not changed or the fish do not become stressed.

E. Projected daily length increase

$$= \frac{\text{expected TU/month}}{\text{TU/in of growth}} \div 30 \text{ days}$$

F. Follow I-B above.

Growth

Depending largely on diet and temperature, marketable size trout (20 to 35 cm in the United States) can be produced in 7 to 14 months of intensive culture, starting with the egg. A few culturists specialize in rearing 2.5-to 16.5-cm fingerlings for sale to other trout farmers. Growth is fastest during the fingerling period, thus such culturists may produce salable fish within 1 to 8 months.

Diseases

Due to the relatively long history of salmonid culture, a comparatively large amount is known about the diseases of these fishes. However, there has been too much emphasis on cure and not enough on prevention. Further, money has literally been thrown down the drain by the needless application of medication. The following hygienic measures should be applied by every culturist:

1. Limiting shipping of live trout at all stages. Not only is there the possibility of infecting one's own stock with someone else's disease or contaminating a disease-resistant strain, but diseases may be introduced to new parts of the world. A number of diseases which create problems for American trout culturists are virtually unknown in Europe and vice versa.
2. If trout are purchased from another culturist, demand a pathological inspection of the stock.
3. Quarantine all diseased fish in ponds which are not connected with enclosures containing healthy fish.
4. Use disinfectants on ponds, nets, and all equipment exposed to disease.

There is still much to be learned about the diseases of salmonids and, while various government agencies are doing their share, it would behave the large commercial enterprises to involve themselves in pathology research. No one is better equipped to understand the problems of trout culture than the culture himself.

Harvest and Processing

As mentioned, there is a considerable market for trout eggs and some opportunity for sale of fingerlings, but the primary product of American trout culture is edible size trout. Some of these are sold for stocking in fish-out ponds, where the public is allowed to angle for them for a fee, but most are processed and marketed iced or frozen. Canadian trout farmers are also considering offering smoked trout.

Small farmers may sell their crop to a central processor, but the large enterprises usually do their own processing. The most highly mechanized operation of this sort is Erkins'. After harvest by seining or pond draining, the trout are transported to the processing plant in specially constructed tank trucks with an oxygen circulating system. (Rangen, Inc., has pioneered a double floored tank truck. Since during transport the trout congregate near the bottom, this has doubled the carrying capacity.) Upon entering the plant, the trout are killed by electrocution to minimize lactic acid, graded-by machine-for the last time, and passed on to an Erkins innovation called an "Eviscerator," which can automatically dress up to 56 trout/min, leaving heads and tails intact. After a careful inspection, the fish are prechilled and given a final washing in circulating refrigerated water, boned, packaged in individual polyethylene bags, and frozen at -22°C. No trout spends more than 30 min from pond to freezing room, thus ensuring a high-quality product.

Prospectus

The Snake River Trout Company and other American producers processed 5.0 million kg of trout in 1969. The majority of this came from the northern and western states, but the greatest growth of the industry is currently occurring in the southern states where, although the climate is too warm to support salmonids in streams, there are numerous large, cold, isothermal springs suitable for trout culture.

Table 16.6 Stocking and Feeding Schedule for Rainbow, Brown and Brook Trout.

Length of fry (cm)	*Size of pond (m2)*	*Depth of pond (m)*	*Number of trout*	*Feed*
Less than 2.5 (sac fry)	1.4	0.3	-	None

2.5-3.0	1.4	0.3	-.	Fresh animal spleen and liver crushed into a juice; earthworms and *Gammarus* may also be fed.
3-5	1.4	0.3	-	Cattle liver-500/0, meat or fish powder-200/0, chopped fish-l 00/0, live chrysalids-200/0; all ingredients chopped and kneaded into a salt mass.
5-7	42	0.4	10,000	Chopped fish-400/0, live chrysalids-400/0, cattle liver-200/0; scattered on surface.
7-9	80	0.65	10,000	Flour, rice bran, and greens-200/0, various animal foods-800/0.
9-15	210	1.0	10,000	Flour, rice, bran, and greens-200/0, various animal foods-800/0.
15-20	400	1.3	10,000	Flour, rice bran, and greens-250/0, various animal foods-750/0.
20-25	400	1.3	400	Flour, rice bran, and greens-250/0, various animal foods-750/0.
25-30	400	1.3	200	Flour, rice bran, and greens-250/0, various animal foods-750/0.
30-40	400	1.3	100	Flour, rice bran, and greens-250/0, various animal foods-75%.

Two factors impede further growth of commercial trout culture in the United States. One is the lack of aggressive and imaginative marketing which has long characterized the American fishery and fish culture industries. The other is competition from other countries which are able to supply trout to the American market cheaper than domestic producers. Most notable in this respect are Denmark and Japan, the only two countries which produce more trout than the United States 10.0 million kg and 5.1 million kg in 1998 and 1999, respectively).

Commercial Trout Culture in Freshwater in Japan

Culture of rainbow, brown, and brook trout in Japan in most respects does not differ essentially from that practiced in the United States. However, fry are kept in shallow running water ponds with a layer of pebbles on the bottom. All life stages receive a diet different from that fed elsewhere.

Diets fed to larger trout vary, but the optimum food is considered to contain 60% protein, 25% fat, 10% carbohydrate, 5% minerals and a variety of vitamins. The diet of trout of all ages is supplemented by live insects lured into the ponds at night by lamps.

COMMERCIAL TROUT CULTURE

Advantages enjoyed Culturists

In some contines is not favoured with an abundance of isothermal springs, so the efficiency of Danish trout culture can never approach that of the best American farms, but Danish culturists do enjoy a number of advantages:

1. They are so far able to operate with earthen ponds and not suffer severe disease problems. Thus construction costs are far less than in the United States. Danish culturists also assert that earthen ponds produce tastier trout.
2. The nearby Baltic Sea and North Sea fisheries provide an excellent source of cheap food in the form of trash fish.
3. The proximity of trout farms, seaports, and research stations in a small country with excellent roads minimizes transportation expenses.
4. Other industries and the government have taken an active interest in trout culture.

Water Supply

Due to the scarcity of spring water, Danish trout culturists are forced to rely upon surface water. Unpolluted streams are abundant only in one area of the country, Jutland, and virtually all of the 600 or so Danish trout farms are located there. The produce of these farms includes live and iced trout for sale in Europe, frozen fish for overseas export, and canned trout. Over 90% of the production of edible size trout is exported. Not all farms include hatchery facilities and some of the smaller operations specialize in producing yed eggs, fry, and fingerlings. The majority of the eyed eggs exported are brown trout, but 85 to 90% of the edible size fish produced are rainbow trout.

Hatchery Techniques and Fry Rearing

Hatchery techniques are essentially the same as those practiced in the United States. Unlike fingerlings and adults, fry are kept in concrete, plastic, or fiberglass tanks up to the size of 5 cm. If kept in earthen ponds, fry are susceptible to "whirling disease," caused by the sporozoan *Lentospora cerebralis*. Young trout are not kept in this sort of container past the fry stage because it is believed to render them more susceptible to fin rot.

Farm Layout and Pond Construction

A typical Danish trout farm might be constructed as shown in Fig. 16.1. A stream is dammed (the culturist is legally responsible for any damage to the stream bank or to wild fish stocks) and diverted into a series of ponds (an average farm would consist of 35 to 60 ponds rather than the 18 shown here), each 30 m long x 10 to 12 m wide, connected to a central outlet channel. Production area is increased by stocking trout in the outlet channel, which is screened off from the stream. The danger of disease is lessened by stocking the ponds at densities considerably lower than those used in the United States.

Feeding

Trout in ponds are fed mainly on minced fish, usually herring and sand eels. If herring are used for a prolonged period of time, the culturist must occasionally add a little vitamin B_1 (thiamine hydrochloride) to counteract the thiaminase in herring, which breaks down vitamin B_1. Fry in tanks are fed a dry food which, although more expensive than fresh fish, is made in Denmark using Danish fish meal and is still relatively cheap. Increasing use is being made of pellets for feeding fingerlings and adults as well. The trout are fed as much as they will eat as long as the water temperature remains below 20°C.

Growth

Conversion ratios are high-5 : 1 to 7 : 1- and growth is slow by American standards. Fry hatched in March or April will reach 8 to 15 cm by November, but then growth practically ceases for the winter. The amount of time required to reach marketable size varies greatly as the 15 or more countries to which Denmark exports differ greatly in the size of trout preferred, but virtually no trout younger than 14 months are marketed and some must be retained for 2 years before being sold.

Trout Farmers' Association

Danish trout farmers maintain direct ties with trout research through a research station established by association of trout farmers. Initially the farmers sustained all the costs, but now the government pays the wages of the sciehtific staff. Opinions differ as to the value of the association, to which most of the nation's trout culturists belong. One of its principal functions is to diagnose and treat diseases, and many farmers undoubtedly find it comforting to have a trained biologist at their beck and call to attend to sick fish. However, some observers have suggested that the time spent running back and forth to fish farms might better be spent in research aimed at increasing production.

Marketing and Prospectus

One thing the association does not do is act as a sales organization. Another group has been formed for this purpose but only 35% of the farmers belong. The rest market their fish through various export firms. Better marketing organization might benefit the Danish trout culture industry, but only if production is increased. Increased production will come only as a result of improved techniques or expansion into brackish and salt water, since virtually all the suitable freshwater is being used.

TROUT CULTURE IN BRACKISH AND SALT WATER

Brackish and Saltwater Trout Farms in Denmark

Both the Japanese and the Danes the limited supplies of freshwater available for trout farming so it is not surprising that these two countries have pioneered culture of these extremely euryhaline fishes in brackish and salt waters. Already there are eight productive Danish trout farms located on a fjord where salinities reach 10% or more, as well as one which uses sea water of about 30% salinity. The water supply for these farms is driven through the ponds by a turbine or propeller. The low sea temperatures in winter raise the danger of mortality, so that freshwater must be used.

Saltwater Trout Culture in Japan

The warmer waters of Japan have permitted one farm, located on Okachi Bay, Miyagi Prefecture, to take the process one step further. One-year-old rainbow trout about 23 cm long are acclimated to sea water, stocked in huge floating net cages anchored about 150 m from shore, and fed on pellets containing fish meal, starch, and vitamins. At the end of about 9 months they have reached a length of 40 cm and are marketed. These trout of course derive a great deal of nourishment from the sea, so that conversion ratios are actually less than 1 : 1. Net cages are now also being employed to grow brown trout and rainbow trout off Tasmania and Norway, Atlantic salmon and rainbow trout off Scotland, and have seen successful experimental use in freshwater with brook trout in Quebec.

LOW-INTENSITY TROUT CULTURE

Estuarine Culture in Denmark

Most brackish and saltwater trout culture schemes are less intensive than those just described. One of the most successful is that carried out by Danish Ministry of Fisheries which, in 1963, with the financial support of the Danish Sport Fishermen's Association, began to release preadapted two-year-old brown trout into fjords. It was hoped that the trout would remain in the fjords for a year or so, fattening themselves for market with no further input of money or labour and be catchable in sufficient quantities to pay for the operation.

The initial releases took place in eight fjords of varying configurations. In three which were relatively closed and sheltered the trout for the most part remained within 12 km of the release site, grew very rapidly, and were recaptured a year later by commercial fishermen in large enough numbers to yield better than 100% profit over the cost of rearing and stocking. Ecological research continues with the aim of determining what sort of fjord is most conducive to growing brown trout, but it has already been estimated that 25% of the 220,000 ha of Danish estuarine waters are suitable for this sort of culture.

Experiments are also being conducted to determine the feasibility of stocking one-year-old fish. Mortality would surely be higher, but rearing costs would be reduced. Growth would also be enhanced, since two-year-old fish reach sexual maturity during their year in the sea. Thus much energy that might have gone into growth is diverted to maturation. By stocking one-year-olds and harvesting just before the onset of maturation this problem could be circumvented.

Saltwater Culture in Japan

Japanese biologists have tested a more sophisticated version of the Danish technique which might be applicable in a wider variety of waters. Rainbow trout, while being acclimated to sea water, were trained to associate the sound of a buzzer with feeding at certain hours. When the trout were acclimated to full strength sea water they were placed in the sea in a net cage off Takamatsu on the island of Shikoku. Training was continued, using an automatic feeding device. When they were thoroughly trained, the net cage was removed and the trout set free to roam the sea at will. However, they continued to return daily to be fed. Reinforcement of the learned behaviour was continued until the trout were harvested.

Using this technique it was possible to harvest 1.2-kg fish after a year and 2.0-to 2.5-kg fish after 2 years. The fish used in the experiments were a special fast-growing strain developed at the University of Washington. (Details on selective breeding of trout follow.) Nevertheless, it is truly remarkable to achieve such growth using inexpensive methods. The process might be intensified by training different species, strains, or age groups of fish to feed at different times of day and to respond to different wavelengths of sound.

Freshwater Culture in Canada

Low-intensity methods may also be effective in freshwater. For example, in Manitoba and Saskatchewan, wheat farmers, who have long unsuccessfully tried to drain the many small, shallow, unproductive lakes, or "potholes," which dot that part of Canada, now stock them with rainbow trout. Fry purchased from hatcheries and stocked in the spring exhibit survival rates of up to 69% and average about 28 cm in length when harvested in the fall. Supplementary feeding is not necessary, since the trout nourish themselves on the abundant crustacean *Gammarus lacustris*. Predation is no problem, since the lakes freeze solid in winter, and thus support no permanent fish populations. The dollar return of this extremely simple form of fish culture is about two and one-half times as much per hectare as the return to wheat farming. Total trout production in the region is expected to eventually increase to 4.5 million fish/year.

CULTURE OF ANADROMOUS SALMONIDS

To Supplement Fisheries

As mentioned, all salmonids are highly euryhaline, and most, if not all species exhibit anadromous traits wherever they have access to the sea. By far the most important anadromous salmonids from the point of view of commercial fisheries or fish culture are the Pacific salmons (genus *Oncorhynchus)*, but some anadromous species of *Salmo, Salvelinus,* and *Thymallus* are also important as human food. By far the most studied of these, due primarily to their great popularity with sport fishermen, are the rainbow trout, the anadromous form of which is often referred to as "steelhead," and the Atlantic salmon. Both are extensively propagated in hatcheries and released to augment sport fisheries. There is at present no commercial fishery for steelhead, but Atlantic salmon are fished commercially in Europe and are propagated for that purpose in several countries. The greatest success in culture of this most difficult of all salmonid species has been achieved in Iceland and Sweden. In Iceland 7.6% of stocked smolts (young salmon about to migrate to sea) return to the fishery, averaging about 7.6 kg in weight. Until recently the Swedish fishery was threatened by the damming of rivers to produce hydroelectric power. Now the fishery is stabilized and at least 20% of the catch is derived from hatchery-raised smolts. Salmon hatched in Swedish hatcheries also contribute to Danish and Finnish fisheries.

A large part of the success of the Swedish hatchery system is attributed to the development by Aktiebolaget Ewos, a Swedish subsidiary of Astra Pharmaceutical Products, Inc., of a special dry food (Table 16.7) acceptable to Atlantic salmon, which usually refuse conventional trout pellets. This diet, which incorporates fish protein concentrate (FPC), has also been found suitable by the Maine Department of Inland Fisheries and Game for rearing the landlocked strain of Atlantic salmon, as well as lake trout, which present similar feeding problems.

Recently Poland has imported eyed Atlantic salmon eggs from Canada for restocking rivers where the species has been exterminated. A trout hatchery is also being built on the Vistula River, where a commercially important run of anadromous brown trout is threatened by dams and pollution.

In Japan, the Dolly Varden *(Salvelinus malma)*, a highly

anadromous species usually scorned by sport fishermen, is raised in hatcheries and stocked in rivers to support a commercial fishery.

The grayling *(Thymallus thymallus)* is generally restricted to wilderness waters, but in Sweden, where grayling spawning rivers have been obstructed by power dams, they are bred in hatcheries and stocked.

More Intensive Culture of Steelheads and Atlantic Salmon

A refinement of the hatchery method of supplementing fisheries for anadromous salmonids has been suggested by Ichthyological Research Corporation of Palo Alto, California. They propose to rear steelhead in a hatchery situated at the head of a short artificial stream, preadapt 15-to 20cm fingerlings to sea water, and release them in the stream. These fish would theoretically enter the sea, to spend the next 2 to 5 years concentrating nutrients. With sexual maturity they would return to the hatchery where they could be harvested immediately with no competition from sport fishermen and none of the deterioration usually associated with long stays in fresh water. By spawning and releasing fish year-round it might be possible to achieve year-round harvests as well. Sport fishery research in Washington has shown that steelhead runs can be artificially established in the manner proposed with 7 to 14% returns.

More intensive culture of a normally anadromous salmonid species is being explored at the University of Rhode Island, where biologists have succeeded in rearing. Atlantic salmon, spawned in freshwater, from an average weight of 0.04 to 0.30 kg in 6 months, using plastic lined pools containing sea water. Improvement is expected as techniques are perfected.

Culture of Freshwater Salmonids to Supplement Commercial Fisheries

Propagation of salmonids to bolster commercial fisheries is not limited to anadromous species. Lake trout have long been stoked for this purpose in the North American Great Lakes. The program was limited and of doubtful value until the parasitic sea lamprey *(Petromyzon marinus)* virtually eradicated the lake trout in the upper Great Lakes, necessitating a crash lake trout culture lamprey control program. This joint effort by the United States and Canadian governments succeeded in re-establishing the fisheries to some extent. If lake trout populations eventually return to anything approaching

Table 16.7 Composition of a Diet for Atlantic Salmon and Size of Particles to be Fed to Fish of Different Sizes.

	Feed analyses		*Feed grades*			
	Starter feed (types f-48) (%)	*Grower feed (%)*	*Grade and size*	*Particle size (cm)*	*Number of fish per kg*	*Fish length (cm)*
Crude protein	58	47.0	Starter			
Fat	8	4.9	1	0.025-0.075	4,994-2,497	2.5-3.5
Ash	15	10.4	2	0.075-0.15	2,497-1,250	3.3-4.1
Water	7	6.5				
Fiber	1	1.5	Grower			
Carbohydrate	11	29.7	2	0.075-0.15	1,250-332	4.1-7.1
Vitamins[a]	–	–	3	0.15-0.23	333-66	7.1-10.9
Minerals[b]	–	–	4	0.23-0.41	66-23	10.9-15.0

a Vitamin A, vitamin D, vitamin E, vitamin K (menadion), vitamin B, (aneurine), vitamin B_2 (riboflavin) B_6 (pyridoxine), vitamin B_{12} (cyanocobalamin), niacin, calcium pentothenate, vitamin H (biotin), folic acid, inositol, p-Aminobenzoic acid, choline, vitamin C (ascorbic acid).

b The product contains all necessary minerals and trace elements.

their former magnitude, it is doubtful whether the present large-scale hatchery program (more than 6 million eggs are produced in Michigan alone) will continue to be justifiable.

Techniques for hatchery propagation of lake trout are essentially the same as those described earlier in this chapter for other salmonids. Until recently, the major source of spawners was fish taken in nets set near the spawning grounds, but the distinct threat that the sea lamprey would completely eliminate the lake trout necessitated the development of hatchery brood stocks.

The incubation period of lake trout eggs is about 2 months at 8.5°C, 4 months at 4°C, and over 5 months at 2°C. The hatching rate of artificially fertilized eggs averages about 65%.

Newly hatched lake trout fry develop slowly and do not absorb the yolk sac until they are about a month old. Their first food should be finely ground meat, such as beef liver or heart. Later, they may be gradually shifted to a conventional hatchery trout diet.

In the Great Lakes, lake trout usually inhabit depths of 30 to 100 m, but, fortunately, lake trout culturists have not found it necessary to simulate this aspect of the natural habitat. Hatchery ponds should, however, be covered or otherwise shaded, since exposure to strong sunlight may cause cataracts.

Lake trout have been stocked at all ages from the alevin stage to two years. Survival of course increases with age. Experiments in Lake Superior indicate that survival is markedly improved by rearing through the first winter, thus most lake trout are now planted in the spring of their second year, at a length of 10 to 13 cm. Rearing to maturity requires 4½ to 6½ years or more.

Selective Breeding and Hybridization

Of the many experimental techniques being applied in efforts to increase trout production probably none has greater potential than selective breeding. Sport fish hatcheries have long practiced selective breeding, at least in a rudimentary manner, for high fecundity, large egg size, high hatching percentage, rapid growth, early maturity, high temperature tolerance, and disease resistance. To facilitate hatchery operations, selection has also been carried out for time of spawning, so that there now exist stocks of rainbow trout which, without manipulation, will spawn in any month of the year.

In recent years strains have also been developed which have a pronounced tendency to leap out of the water when hooked. These "jumpers" are preferred by sport fishermen and are thus sought by fish-out pond operators, as are albino or golden rainbow trout and other "novelty" strains. Cultured trout are enough removed from wild genotypes that trout, particularly rainbow trout, can take their place alongside the common carp *(Cyprinus carpio)* and a few species of ornamental fish as our only truly domesticated aquatic livestock.

Selection is a never-ending process, but it does not take long to achieve some worthwhile results. For example, the California Department of Fish and Game has been able to:

1. Increase the number of rainbow trout spawning at 2 years of age from 53% to 98% in three generations.
2. More than double the average weight of yearling in five generations.
3. Increase egg production by 2-year-old females fourfold in six generations.

Less attention has been paid to selective breeding for characteristics desirable in commercial culture, although most of the characteristics sought by sport fishery culturists would be advantageous to any trout grower. In recent years, a number of workers have sought to improve the commercial breed of rainbow trout. The most famous result of such work is certainly the "supertrout" developed by one of the pioneers in selective breeding of salmonids, Lauren R. Donaldson of the College of Fisheries, University of Washington. After 38 years of selection for 10 characteristics, Donaldson's strain exhibits superior hatching qualities, grows to a length of 67 cm in 3 years, and is so tolerant of high temperatures and certain pollutants that it can be grown in water one would not normally consider fit for the species.

Danish biologists have also developed fast-growing strains of rainbow trout, some of which surpass Donaldson's fish with respect to egg size. These and other selected strains are now available to culturists throughout the world. It should be emphasized, however, that each of these strains is specifically adapted to certain conditions and that no two environments offer the same conditions. It should therefore be the responsibility of each individual culturist not only to start with the best stock obtainable, but to practice selection so as to improve his own strain. Thus far, only a few of the better farmers have undertaken this task.

It is possible that some of the good results claimed for selective breeding should be attributed to other factors. For example, the role of endocrines in determining egg size is not known. But this sort of uncertainty does not negate the importance of selection as an integral part of any efficient farming operation.

Hybridization does not appear to hold as much promise as selection for commercial trout culture. Sport fishery culturists have produced virtually every possible hybrid of rainbow, brown, brook, cutthroat, and lake trout, and Atlantic salmon, but only one or two have proved useful. The only existing hybrid of interest as a food fish is an intraspecific rainbow trout cross developed by Donaldson. By crossing anadromous and landlocked strains, he was able to produce a fish which consistently returns to spawn at two years of age, whereas only a minority of wild steelhead return that early. This hybrid has been used to establish runs in several coastal streams which did not previously support steelhead.

There are other wild strains of trout which should be investigated by culturists. For instance, the deep water Kamloops strain of rainbow trout *(Salmo gairdneri kamloops)* native to certain lakes in British Columbia consistently outgrows all other strains. Before the decimation of the lake trout in the Great Lakes by the sea lamprey there was a subpopulation of a subspecies known as the siscowet *(Salvelinus namaycush siscowet)* in Lake Superior. These fish had a much higher fat or oil content than ordinary lake trout. In fact, large specimens contained up to nearly 70% oil, perhaps the highest oil content of any fish in the-world. There may also be genetic factors involved in the extremely large size attained by steelhead in a few streams such as British Coumbia's Kispiox River and Alaska's Russian River, or by the Lahontan strain of cutthroat trout *(Salmo clarki henshawi)* found only in two Nevada lakes. Hybridization of these strains with domestic stocks might produce beneficial results.

Trout Culture in Thermal Effluent, Recirculating Water Systems, and Tanks

Although trout are generally considered cold water fish, the development of temperature resistant strains raises the possibility that they might be grown in thermal effluent, particularly in areas where normal winter water temperatures greatly retard growth. Heated water might well be employed in conjunction with a recirculating system. Such a system, using small tanks supplied with 14°C water, is presently being successfully used to rear trout fry and fingerlings but commercial application to date have been infrequent.

One such operation exists in Canada, where Sea Pool Fisheries, located in Clam Bay, near Lake Charlotte, Nova Scotia, plan to annually market 1.8 million kg of rainbow trout, Atlantic salmon, brook trout, chinook salmon *(Oncorhynchus tschawytscha)*, and possibly Arctic char *(Salvelinus alpinus)* reared in a semiclosed system in which over 90% of the water is recycled. The basic techniques of filtration, temperature regulation, and so on, are like those used in Columbia River salmon hatcheries, but the facilities are adapted for growth of marketable size fish (up to 7 kg in the case of chinook salmon). Eggs and young fish are maintained in conventional hatchery facilities, while larger fish inhabit a series of large fiberglass pools clustered in groups of three to ten around single filter pools.

A principal feature of Sea Pool Fisheries is the gradual increase in salinity of the water with the age of the fish. By 8 months, at which age rainbow and brook trout have reached marketable size (about 0.2 kg), they are maintained in full strength sea water. In addition to the enhancement of growth and reduction of disease problems associated with high salinity, the sea water brings in with it many organisms which serve as dietary supplements for the trout.

Another interesting development occurred in 1969, when the United States Bureau of Commercial Fisheries began an aquaculture training program for American Indians of the Lummi Reservation near Belling ham, Washington. Part of the training involves raising of trout and Pacific salmon to the smolt stage in a recirculating system using 90 to 95% recycled water. Water contaminated by raw sewage passes through swimming pool filters and a series of ultraviolet lights at 4 to 10 litres min, then through a chiller at 400 to 500 litres/min so that a constant temperature of 10°C is maintained. Another filter is used to treat water that has circulated through the hatchery once or more. The concentration of ammonia nitrogen is monitored daily to determine the amount of new water that must be fed into the system.

A potentially revolutionary breakthrough in trout culture was recently made by the Pennsylvania Fish Commission at its Benner Spring Research Station at Bellefonte. There biologists have succeeded in rearing 20,000 rainbow trout, weighing at least 2720 kg, in a cylindrical fiberglass tank 5 m high and 2.3 m in diameter with a capacity of 20,640 litres. The use of this tank has resulted in the production of more protein per acre than has any other known food production method. If such devices are eventually applied commercially it will be possible to institute trout culture at springs and other water sources located where terrain, space, soil type, or some other factor makes it impossible to construct ponds or raceways. A further advantage is that both construction and maintenance costs for such tanks are much lower than for conventional hatcheries.

Trout in Polyculture

All of the trout culture systems thus far discussed are monoculture systems. Unlike most instances of single-species culture, trout monoculture is not ecologically inconsistent, since these essentially cold water species normally inhabit relatively sterile environments with limited species associations. Nevertheless, trout do have some potential for use in polyculture, particularly as predators to control excess reproduction of small wild fishes which compete for food with cultured species. Trout are sometimes preferred for such purposes over such traditional pond predators as the northern pike *(Esox lucius)* and the largemouth bass *(Micropterus salmoides)* since trout forage in open water, as well as along shorelines and weed beds.

Trout are most commonly used as predators in conjunction with common carp, particularly in Poland and Czechoslovakia, where l-year-old rainbow trout are stocked in ponds at 1200 to 1500/ ha, or about 10 to 15% of the carp population, and are harvested as 2-year-olds. Growth depends principally on the availability of carp fry and other natural foods. In fertile ponds, trout thus stocked may contribute 20 to 50 kg/ha over and above the normal production of carp. Production may be increased by stocking the orfe *(Leuciscus idus)* and the roach *(Rutilus rutilus)*, which do not compete with carp, as food for the trout.

The application of trout in pond culture may be wider than was previously assumed. Russian biologists have found that, as long as dissolved oxygen concentrations remain above 5 ppm most of the time, rainbow trout stocked at low densities do well at 16 to 18°C, continue to feed at up to 24°C, and can withstand temperatures as high as 28°C for short periods.

In 1964 biologists of the Soviet Union's All-Union Institute of Pond Fisheries conducted an experiment with rainbow trout in a more complex polyculture system. Two fertile ponds containing 4 year classes of common carp, 2 year classes of grass carp *(Ctenopharyngodon idella)*, and 1 year class of silver carp *(Hypophfhalmichthys molitrex)* were stocked with 2 year-old rainbow trout as well. The food niche of the trout remained intact, as they ate aquatic and terrestrial insects. *Daphnia*, carp fry, and frogs, most of which were not utilized by the other ©cspecies. At harvest, the trout contributed 30.8 and 36.5 kg/ha, respectively, or 4.7% and 6.3% of the total fish production of the two ponds. Brown trout have been recommended for similar use in polyculture in the Soviet Union.

It has been reported that the taimen *(Hucho taimen)* and the huchen *(Hucho hucho)* are being investigated for fish cultural purposes in China and Spain, respectively. Details are not available, but these large typically solitary and piscivorous salmonids are probably being considered for use as predators.

Index

D

R

S